Introduction to Transformers for NLP

With the Hugging Face Library and Models to Solve Problems

Shashank Mohan Jain

Apress®

Introduction to Transformers for NLP: With the Hugging Face Library and Models to Solve Problems

Shashank Mohan Jain
Bangalore, India

ISBN-13 (pbk): 978-1-4842-8843-6 ISBN-13 (electronic): 978-1-4842-8844-3
https://doi.org/10.1007/978-1-4842-8844-3

Managing Director, Apress Media LLC: Welmoed Spahr
Acquisitions Editor: Celestin Suresh John
Development Editor: James Markham
Coordinating Editor: Shrikant Vishwakarma

Cover designed by eStudioCalamar

Cover image by and machines on Unsplash (www.unsplash.com)

Distributed to the book trade worldwide by Apress Media, LLC, 1 New York Plaza, New York, NY 10004, U.S.A. Phone 1-800-SPRINGER, fax (201) 348-4505, e-mail orders-ny@springer-sbm.com, or visit www.springeronline.com. Apress Media, LLC is a California LLC and the sole member (owner) is Springer Science + Business Media Finance Inc (SSBM Finance Inc). SSBM Finance Inc is a Delaware corporation.

For information on translations, please e-mail booktranslations@springernature.com; for reprint, paperback, or audio rights, please e-mail bookpermissions@springernature.com.

Apress titles may be purchased in bulk for academic, corporate, or promotional use. eBook versions and licenses are also available for most titles. For more information, reference our Print and eBook Bulk Sales web page at http://www.apress.com/bulk-sales.

Any source code or other supplementary material referenced by the author in this book is available to readers on GitHub (https://github.com/Apress). For more detailed information, please visit http://www.apress.com/source-code.

Printed on acid-free paper

Table of Contents

About the Author

Shashank Mohan Jain has been working in the IT industry for around 22 years mainly in the areas of cloud computing, machine learning, and distributed systems. He has keen interests in virtualization techniques, security, and complex systems. Shashank has many software patents to his name in the area of cloud computing, IoT, and machine learning. He is a speaker at multiple reputed cloud conferences. Shashank holds Sun, Microsoft, and Linux kernel certifications.

About the Technical Reviewer

Akshay Kulkarni is a renowned AI and machine learning evangelist and thought leader. He has consulted several Fortune 500 and global enterprises on driving AI and data science–led strategic transformation. Akshay has rich experience in building and scaling AI and machine learning businesses and creating significant impact. He is currently a data science and AI manager at Publicis Sapient, where he is a part of strategy and transformation interventions through AI. He manages high-priority growth initiatives around data science and works on various artificial intelligence engagements by applying state-of-the-art techniques to this space. Akshay is also a Google Developers Expert in machine learning, a published author of books on NLP and deep learning, and a regular speaker at major AI and data science conferences. In 2019, Akshay was named one of the top "40 under 40 data scientists" in India. In his spare time, he enjoys reading, writing, coding, and mentoring aspiring data scientists. He lives in Bangalore, India, with his family.

Introduction

This book takes the user through the journey of natural language processing starting from n-gram models to neural network architectures like RNN before it moves to the state-of-the-art technology today, which is known as the transformers. The book details out the transformer architecture and mainly explains the self-attention mechanism, which is the foundation of the transformer concept.

The book deals with the topic of transformers in depth with examples from different NLP areas like text generation, sentiment analysis, zero-shot learning, text summarization, etc. The book takes a deep dive into huggingface APIs and their usage to create simple Gradio-based applications. We will delve into details of not only using pretrained models but also how to fine-tune the existing models with our own datasets.

We cover models like BERT, GPT2, T5, etc., and showcase how these models can be used directly to create a different range of applications in the area of natural language processing and understanding.

The book doesn't just limit the knowledge and exploration of transformers to NLP but also covers at a high level how transformers are being used in areas like vision.

Source Code

All source code used in this book can be found at `github.com/apress/intro-transformers-nlp`.

CHAPTER 1

Introduction to Language Models

Language is power, life and the instrument of culture, the instrument of domination and liberation.

—*Angela Carter (English writer)*

One of the biggest developments that made *Homo sapiens* different from other animal species on this planet was the evolution of language. This allowed us to exchange and communicate ideas and thoughts, which led to so many scientific discoveries including the Internet. This is how important language is.

So when we venture into the area of artificial intelligence, the progress made there would not move much unless we made sure the machines understand and comprehend natural language. So it's pertinent for anyone who wants to venture into the area of artificial intelligence, and thereby artificial general intelligence, that they develop a good grasp of how we are progressing on teaching machines how to understand language.

The intent of this chapter is to take you through the evolution of the natural language processing domain by covering some of the historical aspects of it and its evolution to the state-of-the-art neural network–based language models of today.

© Shashank Mohan Jain 2022
S. M. Jain, *Introduction to Transformers for NLP*,
https://doi.org/10.1007/978-1-4842-8844-3_1

History of NLP

Natural language processing is one of the fastest-developing areas of machine learning and artificial intelligence. Its aim is to provide machines the capability to understand natural language and provide capabilities that assist humans in achieving tasks related to natural language. The concept of machine translation (also known as MT), which was first developed during the Second World War, was the seed from which it grew. The intent of NLU, or natural language understanding, is to allow machines to comprehend natural language and accomplish tasks like translation from one language to another, determining the sentiment of a specific text segment, or providing a summary of, say, a paragraph.

A language can be broken down into its component parts, which can be thought of as a set of rules or symbols. Following the integration process, these symbols are put to use for both the transmission and the broadcasting of information. The field of natural language processing is broken down into several subfields, the most notable of which are natural language generation and natural language understanding. These subfields, as their names imply, are concerned with the production of text as well as its comprehension. Be careful not to let these relatively new words – such as phonology, pragmatics, morphology, syntax, and semantics – throw you off.

One of the main areas of NLP or NLU is to understand not just the statistical properties of a particular language but also the semantics of it. With machine learning the aim was to feed in the content in a certain language to the machine and let the machine understand not just the statistical properties but also the meaning and context of, say, a certain word.

An NLP engineer's workflow would consist of first attempting to transform the words into numbers that computers are able to interpret and then developing machine learning architectures that are able to use these numbers for the many tasks that are necessary. In more specific terms, it should involve the following steps:

1. Collecting Data: The first thing to do with every project you have in mind is to collect data that is directly connected to the project you are working on. This is essential to the study of machine learning, which is its own area. We supply many algorithms with vast volumes of data, some of which may have been prohibitively expensive to obtain. In the context of NLP, this stage may entail the collection of tweets or reviews from various ecommerce websites like Amazon or Yelp. Additionally, this process may involve cleaning and categorizing the collected tweets or reviews.

2. Tokenization: This is the process of chopping up each piece of text into manageable word chunks in order to get ready for the subsequent stage. In this phase, you may also be asked to remove stop words, perform lemmatization, or stem the text.

3. Vectorization: In this stage, the tokens that were obtained in the previous step known as "tokenization" are transformed into vectors that the ML algorithms are able to process. It is clear to us at this point that the models that we develop do not truly see the words and comprehend them in the same manner that we humans do (or that we think we do), but rather operate on the vector representations of these words.

4. Model Creation and Evaluation: This step entails developing ML models and architectures, such as transformers, that are able to chew on the word vectors that are provided for the many tasks that are

required. Translation, semantic analysis, named
entity identification, and other similar activities
may be included in these tasks. The NLP engineer
is responsible for performing ongoing evaluations
of the models and architectures, measuring them
against previously established goals and KPIs.

When we want to understand something, we need to form a mental
model of that thing. As an example, if I say I understand how a cat looks
like, I have a mental model of its features like fur, eyes, legs, etc. These are
features or dimensions that enable us to represent a thing or a concept.

Similarly, to begin understanding a sentence or a word, first of all,
we need a mathematical representation of the word itself. The reason
for this is that the machines only understand numbers. So we need an
approach to encode the representation in numbers. We start with a very
simplistic statistical approach called bag of words, which is described in
the following.

Bag of Words

The bag of words is a statistical technique based on word counts and using
that to create mathematical representations of documents in which these
words occur. When we say a mathematical representation, we mean a
mathematical vector, which represents a document in vector space, where
each word can be thought of as a separate dimension in that space.

Let's take a simple example. Assume we have three documents with
one sentence each as illustrated in the following:

Document 1: I am having fun of my lifetime.

Document 2: I am going to visit a tourist
destination this time.

Document 3: Tourist destinations provide such fun.

Now what we do is that we first create a vocabulary based on all the words present in our document set. As our document set constitutes these three documents, we will have a vocabulary as in the following (we only will take the unique words):

I

am

having

fun

of

my

lifetime

going

to

visit

tourist

destination

this

time

provide

such

So we can see we have 16 words here (for simplicity reasons, we are keeping words like *a*, *the*, etc. in the set). This becomes our corpus. Now think of each word as a dimension in 16-dimensional vector space.

If we take document 1, the dimensions can be coded as in the following:

[1,1,1,1,1,1,1,0,0,0,0,0,0,0,0,0]

Document 2 will look like this:

[1,1,0,0,0,0,1,1,1,1,1,1,1,0,0,0]

Document 3 will look like this:

[0,0,0,1,0,0,0,0,0,0,1,1,0,0,1,1]

Here, 1 is representing the presence of a word in the document. This mechanism is also called one-hot encoding and is a most simplistic representation of a document or a sentence. As we will move ahead in the book, we will see how this mechanism of word representation gets better and better.

This representation allows the machine to play around with these numbers and perform mathematical operations on them.

n-grams

Before we start to understand more of Ngrams, we will take a detour and first understand something called the bag of words model.

Because of the sequential quality of language, the sequence in which words are presented in a text is of utmost importance. This is something that we are all well aware of, even if we don't consciously think about it very often. n-grams provide us the capability to predict the next word given the previous words.

The primary concept behind creating text with n-grams is based on the statistical distribution of words. The main idea is to determine the probability of occurrence of the nth word in the sequence given the probability of occurrence of n-1 words prior to that word. In effect it uses the chain rule of probability to predict the occurrence of a word with a certain probability.

This calculation is shown in the following:

$$P\left(x^{(t+1)}\middle|x^{(t)},...,x^{(1)}\right) = P\left(x^{(t+1)}\middle|x^{(t)},..., x^{(t-n+2)}\right)$$

The left side of the equation signifies the probability of seeing a word x(t+1) given that we have seen words from x(1) to x(t).

The whole concept rests on the chain rule of probability, which in simple terms allows us to represent a complex joint distribution to be factored into a set of conditional distributions.

Let's examine a simple illustration of a trigram model. A trigram model keeps the context of the last two words to predict the next word in the sequence.

Take for example the following utterance:

"Anuj is walking on the ___."

Here, we need to predict the word after the utterance "on the."

We make an assumption here that, based on the dataset, these are the following probable continuations: "road," "pavement."

Now we need to calculate the probabilities:

P(road |on the) and P(pavement| on the)

The first probability is for the occurrence of the word *road* given the words *on the* have already occurred before it. The second probability is for the word *pavement* given the words *on the* have occurred before it.

After the probabilities have been calculated, the word that has the highest conditional probability would be the next word.

We can restate as a simple mechnism based on statistical methods to calculate the probability of occurrence of the next word given the context of previous words. As the corpus grows big and the number of sentences increases, doing calculations beyond simple bigrams will be extremely challenging. We need to learn a way to generate these conditional probability distributions. By conditional distribution we mean given some words, we need to understand the probability distribution of words that can occur next. Since we don't know what the shape of these distributions could be, we use neural networks to approximate the parameters of such distributions. In the next section, we will cover recurrent neural networks (RNNs), which allow us to achieve this task.

Recurrent Neural Networks

Siri on Apple products and voice search on Google products both make use of recurrent neural networks (RNNs), the most advanced algorithm currently available for processing sequential input.

As the name suggests, it's a recurrent neural network and due to their recurrent nature, they are best suited for handling sequences of data, like time series or languages. When dealing with sequences, the most important aspect is to handle context, which entails remembering what has happened in previous sequences and using that information to represent the current input in a better way.

Because it is the only algorithm that has an internal memory, recurrent neural networks (RNNs) are extremely strong and reliable. They are also among the most promising algorithms now in use.

In comparison with many other types of deep learning algorithms, recurrent neural networks have been around for quite some time. They were first developed in the 1980s, but it has only been in the most recent decades that we have realized the full extent of their potential.

Though RNNs have been used for handling sequential data, they tend to suffer from certain issues like vanishing gradients, and they are unable to capture long-term dependencies. This has led to the emergence of new neural network architectures like LSTM (Long Short-Term Memory) and GRU, which overcame the issues with RNN.

LSTMs and GRUs, which have their own internal memories and also have more refined architectures than a plain RNN, are able to remember significant aspects of the data input they were given, which enables them to make very accurate forecasts regarding what will happen in the future.

LSTMs and GRUs became the default neural network architectures when it came to handling sequential data like time series or languages or data provided by sensors deployed in the field mainly for IoT-based applications.

What Exactly Is a Recurrent Neural Network (RNN)?

A subcategory of neural networks known as recurrent neural networks, or RNNs, are advantageous for the modeling of sequence data. RNNs are derived from feed-forward networks and display behavior that is analogous to the way in which human brains operate. Put another way, recurrent neural networks are capable of producing predicted outcomes with sequential data, while other algorithms are unable to do so.

How RNNs Work

In order to have a complete comprehension of RNNs, you will need to have a functional understanding of "regular" feed-forward neural networks as well as sequential data.

The most fundamental definition of sequential data is simply ordered data in which related items follow one another in time. Examples of this include the DNA sequence and financial data. The most common kind of sequential data is probably time series data, which is nothing more than a string of data points that are presented in the appropriate chronological sequence.

The manner in which information is transmitted is what gives RNNs and feed-forward neural networks their respective titles.

A feed-forward network is a single-pass network, which means there is no recurrence in it. Input is passed via various layers, and the output of the network is compared with actual output, and error correction is introduced via the backpropagation (BP) mechanism. This is only helpful in handling one input at a time.

Because it can just handle one input at a time, a feed-forward neural network is terrible at forecasting what will happen next because it has no memory to store the information that it takes in. A feed-forward network does not have any concept of chronological order because it simply takes into account the most recent input. It simply has no memory of anything that occurred in the past, other than the training that it received.

In an RNN, the information is repeated endlessly within a loop. When it comes time to make a choice, it takes into account the most recent input in addition to the lessons it has picked up from the inputs it has previously been given.

Figure 1-1 presents this difference in a visual form.

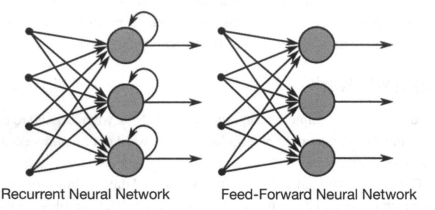

Recurrent Neural Network   Feed-Forward Neural Network

Figure 1-1. *Difference between a simple feed-forward network and an RNN*

Another effective method of illuminating the idea of the memory of a recurrent neural network is to describe it by way of an illustration.

Imagine you have a typical feed-forward neural network and you feed it the word "machine" as an input. You then watch as the network processes the word, one character at a time. Because by the time it reaches the character "h," it has already forgotten about "m," "a," and "c," it is nearly difficult for this form of neural network to anticipate which character will come next because it has already forgotten about those characters.

However, due to the fact that it possesses its own internal memory, a recurrent neural network is able to remember those characters. It generates output, replicates that output, and then feeds both of those versions back into the network.

To put it another way, recurrent neural networks incorporate recent history into the analysis of the present.

As a result, an RNN takes into consideration both the present and the recent past as its inputs. This is significant because the sequence of data provides essential information about what is to come after it. This is one of the reasons an RNN is able to accomplish tasks that other algorithms are unable to.

The output is first produced by a feed-forward neural network, which, like all other deep learning algorithms, begins by assigning a weight matrix to the inputs of the network. Take note that RNNs assign weights not only to the most recent input but also to the inputs that came before it. In addition, a recurrent neural network will adjust the weights over the course of time using gradient descent as well as backpropagation (BP).

As things evolved, we discovered that there are challenges with RNNs and their counterparts in terms of processing time and capturing long-term dependencies between words in a sentence. This led us to the evolution of language models, which we describe in the next section.

Language Models

Over the course of the last decade, there has been a substantial amount of development in the field of information extraction from textual data. Natural language processing replaced text mining as the name of this field, and as a result, the approach that is applied in this field has also undergone a significant change. The development of language models as a foundation for a wide variety of applications that seek to extract useful insights from unprocessed text was one of the primary factors that brought about this transition.

A probability distribution over words or word sequences is the fundamental building block of a language model. In application, a language model provides the chance that a particular word sequence can be considered "valid." In this discussion, "validity" does not at all refer to the grammatical correctness of a statement. This indicates that it is similar to the way people speak (or, to be more specific, write) because this is how

11

the language model acquires its knowledge. A language model is "just" a tool to incorporate abundant information in a condensed manner that is reusable in an out-of-sample setting. This is an important point to keep in mind because it shows that there is no magic to a language model (like other machine learning models, particularly deep neural networks).

What Advantages Does Using a Language Model Give Us?

The abstract comprehension of natural language, which is required in order to deduce word probabilities based on context, can be put to use in a variety of contexts and activities.

We are able to execute extractive or abstractive summarization of texts if we have an accurate language model. If we have models for a variety of languages, it will be much simpler to develop a system for automatic translation. Among the more complicated applications is the process of question answering (with or without context). Language models today are being put to very interesting tasks like software code generation, text-to-image generation like DALL-E 2 from OpenAI, as well as other text generation mechanisms like GPT3 and so on.

It is essential to understand that there is a distinction between

a)   Statistical techniques that come under probabilistic language models

b)   Language models that are built on neural networks

As explained in the "n-grams" section, calculating the probabilities of n-grams results in the construction of a straightforward probabilistic language model (a) (an n-gram being an n-word sequence, n being an integer greater than 0). The likelihood of an n-gram can be defined as the conditional probability that the n-gram's final word follows a specific n-1 gram (leaving out the last word). In everyday terms, it refers to the frequency of occurrence of the final word that comes after the n-1 gram that does not include the last word. Given the n-1 gram, which represents the present, the probabilities of the n-gram, which represents the future, do not depend on the n-2, n-3, etc. grams, which represent the past. This is an assumption made by Markov.

There are obvious disadvantages associated with taking this method. The probability distribution of the following word is only influenced by the n words that came before it. This is the most essential point. Texts that are difficult to understand contain rich contexts, which can have a significant impact on the choice of the next word. Therefore, the identity of the following word might not be discernible from the n words that came before it, even if n is as high as 50.

On top of that, it is obvious that this method does not scale well: the number of possible permutations skyrockets as the size of the dataset (n) increases, despite the fact that the majority of the variants never appear in the actual text. In addition, each and every occurring probability (or each and every n-gram count) needs to be computed and saved! Additionally, non-occurring n-grams cause a sparsity problem because the granularity of the probability distribution might be relatively low when there are so few of them (word probabilities have few different values; therefore, most of the words have the same probability).

Neural Network–Based Language Models

The manner in which neural network–based language models encode inputs makes it easier to deal with the sparsity problem. Embedding layers produce a vector of arbitrary size for each word, where the semantic links between the words is taken into account. These continuous vectors generate the granularity that is so desperately required in the probability distribution of the following word. In addition, the language model is basically a function (as are all neural networks, which include a great deal of matrix calculations), which means that it is not necessary to store all of the n-gram counts in order to construct the probability distribution of the next word.

The sparsity problem can be solved using neural networks, but there is still a difficulty with the context. First, the process of developing language models consisted of finding ways to solve the context problem in an ever-increasingly effective manner. This was accomplished by bringing in ever-more context words in order to impact the probability distribution in an ever-more effective manner. Second, the objective was to design an architecture that would endow the model with the capability of discovering which phrases in a given context are more significant than others.

Utilizing recurrent neural networks (described in the previous section) is a step in the right direction with regard to the topic of handling context. When selecting the following word, it takes into account all of the words that came before it because it is either an LSTM or a GRU cell-based network.

The fact that RNN-based designs are sequential is the primary downside associated with using these kinds of models. Due to the absence of any opportunity for parallel processing, the amount of time required for training skyrockets when dealing with lengthy sequences. The transformer architecture is the answer to this predicament that you're having. It is recommended that you read the original document that was produced by Google Brain.

Additionally, OpenAI's GPT models and Google's BERT are utilizing the transformer design (which we will discuss in upcoming chapters). In addition to this, these models make use of a technique known as attention, which allows the model to discover which inputs, in particular circumstances, merit more attention than others.

In terms of the architecture of models, the most significant quantum leaps were firstly made by RNNs (specifically LSTM and GRU), which solved the sparsity problem and allowed language models to use a great deal less disk space, and then, more recently, by the transformer architecture, which made parallelization possible and created attention mechanisms. Both of these developments were important. However, architecture is not the only domain in which a language model can demonstrate its prowess.

OpenAI released few language models based on the transformer architecture. The first model was GPT1, and the latest one is GPT3. When contrasted with the architecture of GPT1, GPT3 contains almost no innovative features. But it is massive. It was trained using the largest corpus a model has ever been trained on, which is known as the Common Crawl, and it has 175 billion different parameters. This is made possible in part by the semi-supervised training method of a language model, which allows for the use of a text as a training example while simultaneously omitting some words. The remarkable power of GPT3 stems from the fact that it read virtually all of the text that has been published anywhere on the Internet over the course of the last few years and possesses the capacity to reflect the majority of the complexity that natural language possesses.

In conclusion, I would like to present the T5 model that Google offers (Figure 1-2). In the past, language models were utilized for conventional natural language processing tasks, such as part-of-speech (POS) tagging or machine translation, albeit with a few tweaks here and there. Because of its abstract capability to understand the fundamental structure of natural language, BERT, for instance, may be retrained to function as a POS tagger with only a little bit of additional instruction.

Figure 1-2 shows a representation of the T5 language model.

When using T5, there is no requirement for any sort of modification in order to complete NLP jobs. If it receives a text that contains certain M

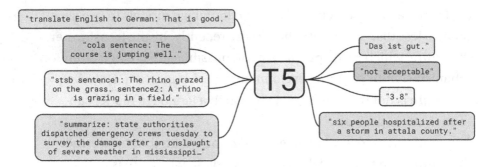

Figure 1-2. *Google's T5 model, which is a text-to-text model*

tokens, it recognizes that these tokens represent blanks that need to be filled in with the proper words. It is also capable of providing answers. If it is provided with some context after the questions, it will seek that context for the answer; otherwise, it will answer based on the information it already possesses. It is an interesting fact that its designers have been bested by it in a trivia contest. Additional examples of possible applications can be seen in the image (Figure 1-2) on the left.

Summary

In my opinion, NLP is the area in which we are the most likely to succeed in developing artificial intelligence. There is a great deal of excitement about this term, and many straightforward decision-making systems, as well as virtually every neural network, are referred to as AI; however, this is primarily marketing. The Oxford Dictionary of English, as well as just about any other dictionary, defines *artificial intelligence* as the performance of intelligence-related tasks by a machine that are analogous to those performed by humans. One of the key aspects for AI is generalization, which is the ability of a single model to do many tasks. The fact that the same model can perform a wide variety of NLP tasks and can infer what to do based on the input is astounding in and of itself, and it gets us one step closer to genuinely developing artificial intelligence systems that are comparable to human intellect.

CHAPTER 2

Introduction to Transformers

Around December 2017, a seminal paper titled "Attention Is All You Need" was published. This paper revolutionized the way the NLP domain would look like in the times to come. Transformers and what is known as a sequence-to-sequence architecture are described in this paper.

A sequence-to-sequence (or Seq2Seq) neural network is a neural network that converts one sequence of components into another, such as the words in a phrase. (Considering the name, this should come as no surprise.)

Seq2Seq models excel at translation, which involves transforming a sequence of words from one language into a sequence of other words in another. Long Short-Term Memory (LSTM)–based neural network architectures are deemed best for the sequence-to-sequence kind of requirements. The LSTM model has a concept of forget gates via which it can also forget information, which it doesn't need to remember.

© Shashank Mohan Jain 2022
S. M. Jain, *Introduction to Transformers for NLP*,
https://doi.org/10.1007/978-1-4842-8844-3_2

What Is a Seq2Seq Neural Network?

A Seq2Seq model is one that begins with a sequence of objects (such as words, letters, or time series) and produces still another sequence of items as its output. When it comes to neural machine translation, we need to provide an input sentence in a specific language, and output should be the translated text in another language.

As shown in Figure 2-1 , the neural network based on the Seq2Seq architecture takes the word *learn* as input and outputs the French translation of the word.

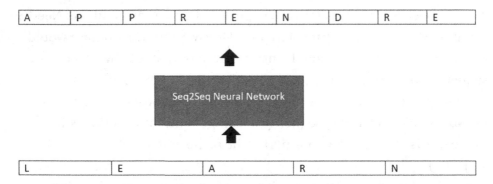

Figure 2-1. *Function of a Seq2Seq network*

The encoder and decoder are the two components that make up the model as shown in Figure 2-2. The input sequence's context is saved by the encoder in the form of a vector, which it then transmits to the decoder so that the decoder may construct the output sequence based on the information it contains.

The architecture of encoders and decoders was mainly manifested by RNNs, LSTMs, or GRUs for sequence-to-sequence tasks.

Figure 2-2 explains how the encoder and decoder fit into the Seq2Seq architecture for a translation task.

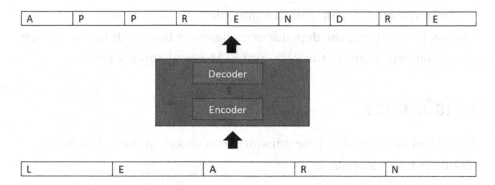

A	P	P	R	E	N	D	R	E

L	E	A	R	N

Figure 2-2. *Encoder-decoder usage in Seq2Seq networks*

We won't go into details of the Seq2Seq architecture since the book is mostly about transformers. But to understand the essence of Seq2Seq models, they take a sequence as input and transform it into another sequence.

The Transformer

As was discussed in the beginning of this chapter, there was this great paper called "Attention Is All You Need" in which a new neural architecture called the transformer was proposed. The main highlight of this work was a mechanism called self-attention. One of the idea transformer architecture was to move away from sequential processing where we provided input to the network one at a time. For example, in the case of a sentence, the RNN or LSTM will take one word input at a time from the sentence. This would mean the processing will only be sequential. Transformers intended to change this design by providing the whole sequence as input at one shot to the network and allowing the network to learn one whole sentence at once. This would allow parallel processing to happen and also allow distributing the learnings to other cores or GPUs in parallel.

21

The transformer architecture's main point was to only use self-attention for capturing the dependencies between the words in a sequence and not depend on any of the RNN- or LSTM-based approaches.

Transformers

A high-level architecture of the transformer is shown in the following. It constitutes two major parts:

1. Encoder

2. Decoder

Figure 2-3 shows the transformer architecture with both encoder and decoder blocks.

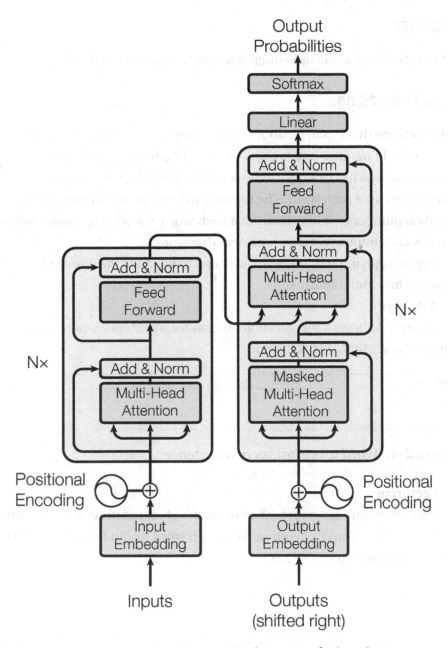

Figure 2-3. *High-level transformer architecture (taken from Vaswani's "Attention Is All You Need" paper)*

Encoder

Let us take a look at the individual layers of the encoder in detail.

Input Embeddings

The first thing that has to be done is to feed input into a layer that does word embedding. One way to think of a word embedding layer is as a lookup table, which allows for the acquisition of a learned vector representation of each word. The learning process in neural networks is based on numbers, which means that each word is associated with a vector that has continuous values to represent that word.

Before applying self-attention, we need to tokenize the words of a sentence into individual tokens.

Tokenize

Figure 2-4 shows a tokenization process for a sentence where each word of the sentence is tokenized.

Working	Whole	Day
↓	↓	↓
T1	T2	T3

Figure 2-4. *Word tokenization in a sentence*

Vectorize

Post tokenization we vectorize the tokens obtained previously by using, say, word2vec or GloVe encoding. Now each word would be represented as a vector as shown in Figure 2-5.

	Working	Whole	Day
Tokenize	↓ T1	↓ T2	↓ T3
Vectorize	↓ V1	↓ V2	↓ V3

Figure 2-5. *Vectorize each token individually to create a mathematical representation of the word*

Positional Encoding

It's an encoding based on the position of a specific word, say, in a sentence.

This encoding is based on which position the text is within a sequence. We need to provide some information about the positions in the input embeddings because the transformer encoder does not have recurrence as recurrent neural networks do. Positional encoding is what is used to accomplish this goal. The authors conceived of a cunning strategy that makes use of the sine and cosine functions.

The mathematical representation for positional encoding is shown in Equation 2-1:

$$PE_{(pos,2i)} = sin\left(pos / 10000^{2i/d_{model}} \right)$$
$$PE_{(pos,2i+1)} = cos\left(pos / 10000^{2i/d_{model}} \right)$$

Equation 2-1

This equation calculates the positional embeddings of each word so that each word has a positional component alongside the semantic component.

The mathematical specifics of positional encoding are outside the scope of this discussion; however, the fundamentals are as follows.

As demonstrated in Equation 2-1, we pass the word at the odd index via the cosine function and at the even index via the sine function. The output of each word would be a vector that represents the positional encoding of the specific word. We explain this in the following via an example.

Example of Positional Encoding

Take a sequence, say, "Sam loves to walk."

Here, first we define the values of some parameters:

N: 10000.

K: Position of a word in a sentence starting from 0.

D: Dimension of a sentence. In our case it's 4.

I: Used for mapping to column index.

Now we calculate the values of the sine and cosine components.

Sam

$\sin(0/10000^{(0/4)})=0$	$\cos(0/10000^{(0/4)})=1$	$\sin(0/10000^{(2/4)})=0$	$\cos(0/10000^{(2/4)})=1$

Likes

$\sin(1/10000^{(0/4)})=$ 0.8414	$\cos(1/10000^{(0/4)})=$ 0.5403	$\sin(1/10000^{(2/4)})=$ 0.0099	$\cos(1/10000^{(2/4)})=$ 0.999

To

$\sin(2/10000^{(0/4)})=$ 0.909	$\cos(2/10000^{(0/4)})=$ -0.416	$\sin(2/10000^{(2/4)})=$ 0.0199	$\cos(2/10000^{(2/4)})=$ 0.9998

Walk

$\sin(3/10000^{(0/4)})=$ 0.1411	$\cos(3/10000^{(0/4)})=$ -0.989	$\sin(3/10000^{(2/4)})=$ 0.0299	$\cos(3/10000^{(2/4)})=$ 0.9995

We can see how each word from the sentence "Sam likes to walk" has got a word embedding vector to represent the position of the word in the sentence.

After that, add those vectors to the input embeddings that correspond to those vectors. The intuition behind adding the positional embeddings to the input embeddings is to give each word a small shift in vector space toward the position the word occurs in. So if we think a bit more about it, we can see this leads to semantically similar words that occur in near positions to be represented close by in vector space. Because of this, the network is provided with accurate information regarding the location of each vector. Because of their linear characteristics and the ease with which the model may learn to pay attention to them, the sine and cosine functions were selected together.

At this point, we have reached the encoder layer, as also shown in Figure 2-6. The objective of the encoder layer is to convert all the input sequences into a representation that captures the context in a way that it also gives more attention to words that are more important to it in a specific context. It begins with a multi-headed attention sub-module and then moves on to a fully connected network sub-module. Finally, this is followed by a layer of normalization.

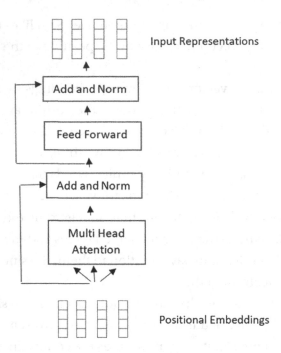

Figure 2-6. *Encoder component of the transformer*

Multi-headed Attention

The encoder makes use of a particular attention process known as self-attention in its multi-headed attention system. A multi-headed attention is nothing but multiple modules of self-attention capturing different kinds of attentions. As specified earlier in the chapter, self-attention allows us to associate each word in the input with other words in the same sentence.

Let's go a bit deeper into the self-attention layer.

Take this sentence for example:

"Sam had been working whole day so he decided to take some rest"

In the preceding sentence, as humans when the word *he* is uttered, we automatically connect it to *Sam*. This simply implies that when we refer to the word *he*, we need to attend to the word *Sam*. The mechanism of self-attention allows us to provide a reference to *Sam* when *he* is uttered in this sentence. As the model gets trained, it looks at each individual word in the input sequence and tries to look at words it must attend to in order to provide a better embedding for the word. In essence it captures the context better by using the attention mechanism.

One can see in Figure 2-7 that the word *he* pays attention to the word *Sam*. By using the self-attention technique, the context gets captured in a much better way.

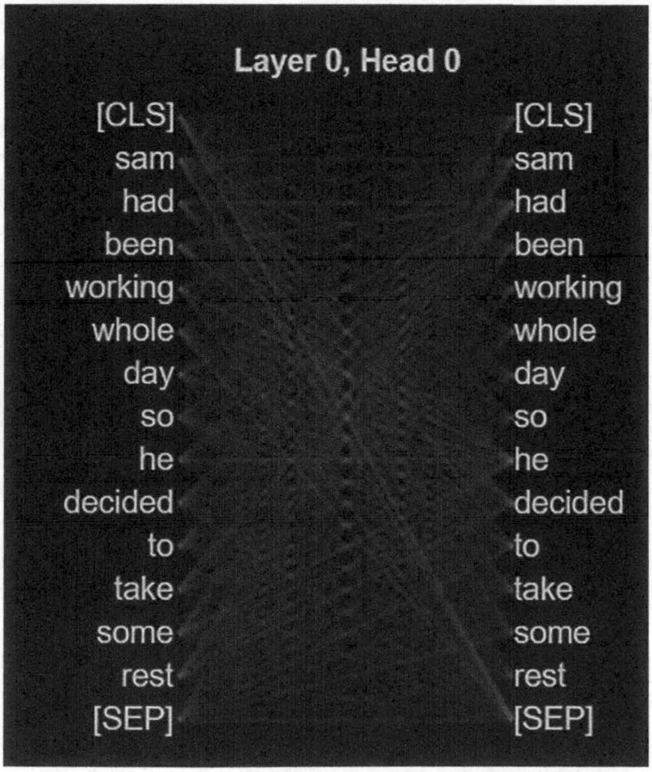

Figure 2-7. Attention mechanism at work

Next, we take a look into how query, key, and value vectors work.

Query, Key, and Value Vectors

The concept of self-attention is based on three vector representations:

1. Query

2. Key

3. Value

The query, key, and value concept emerged from information retrieval systems. As an example when we search in Google or for that matter any search engine, we provide a **query** as input. The search engine then finds specific **keys** in its database or repository that match the query and finally will return the best **values** back to us as a result.

Figure 2-8 shows how the query, key, and value mechanism works to capture context information about a word representation.

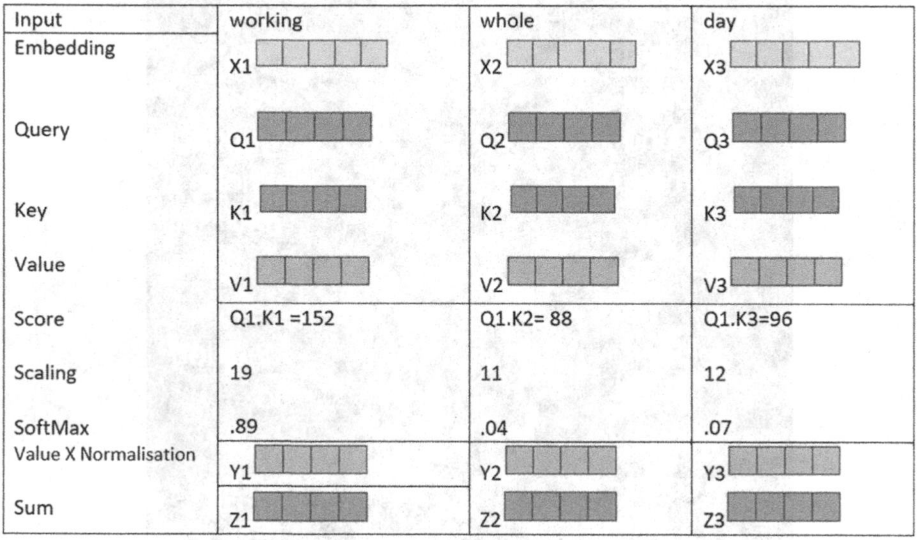

Input	working	whole	day
Embedding	X1	X2	X3
Query	Q1	Q2	Q3
Key	K1	K2	K3
Value	V1	V2	V3
Score	Q1.K1 =152	Q1.K2= 88	Q1.K3=96
Scaling	19	11	12
SoftMax	.89	.04	.07
Value X Normalisation	Y1	Y2	Y3
Sum	Z1	Z2	Z3

Figure 2-8. *Query, key, and value mechanism at work*

Self-Attention

As explained in Figure 2-8, once we have the input vector prepared, we pass it to a self-attention layer, which actually captures the context of the word in its representation by using an attention mechanism.

Figure 2-9 shows how the self-attention mechanism works to create context vectors.

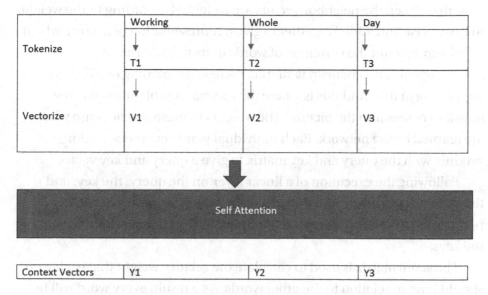

Figure 2-9. *Self-attention mechanism*

For a simple implementation of attention at a high level, we take individual word vectors obtained from the preceding step, and for each word vector

1. We do a dot product with self and other vectors.

2. We get a score, say, S11,S12,S13 – in this case we have three words in a sentence and we want to capture embedding for V1.

3. This score is then treated as weight and multiplied by V1, V2, and V3.

4. Normalize the score.

5. The sum of the vectors obtained in step 3 is done.

The intuition of the preceding steps is to capture the closeness of words in vector space to each other and then assign weights based on that closeness. Then the neighbor vectors are weighted according to the weight they exercise and added together to give a representation of a word, which takes into account the closeness of words in its neighborhood.

Though this mechanism is simple, there is no learning of weights happening in this. And this is where the mechanism of query and key matrices comes into the picture. The weights in these matrices are what are learned by the network. Each individual word vector does a dot product with the query and key matrix to give a query and key vector.

Following the execution of a linear layer on the query, the key, and the value vector, the production of a score matrix is accomplished by performing a dot product matrix multiplication on the queries and the keys.

The score matrix is used to calculate the relative weight that each word should have in relation to the other words. As a result, every word will be assigned a score, keeping the surrounding contextual words. The higher the score for a specific word, the more it is attended to. This is the process that is used to map the queries onto the keys.

After that, the scores are lowered by having their total divided by the square root of the dimension that contains both the query and the key. This is done in order to make it possible to build gradients that are more stable, as multiplying values can have explosive effects.

After that, you must take the softmax of the scaled score in order to obtain the attention weights, which will provide you with probability values ranging from 0 to 1. When you perform a softmax, the better results are amplified, while the lower levels see a downward adjustment. Because of this, the model is able to have a greater sense of confidence regarding the terms to which they should pay attention.

Multiply the Softmax Output with the Value Vector

After that, you get an output vector by taking the attention weights and multiplying them by your value vector. The higher the softmax scores, the more vital it will be to preserve the values of the words that the model learns. The words with lower scores will be drowned out by those with higher scores. The output of it is then processed by a linear layer after it has been fed into it.

Figure 2-10 shows the mechanism of multiplication of attention weights with the value vectors to get the output.

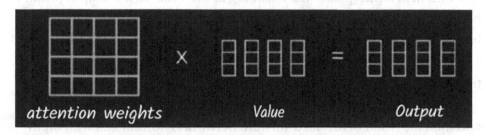

Figure 2-10. Multiply attention weights by the value vectors

Computing Multi-headed Attention

Before applying self-attention, you must first divide the query, the key, and the value into N vectors so that you may do a multi-headed attention computation with this data. After that, each of the split vectors goes through its own particular process of self-attention. The term "head" refers to each individual process of self-attention. Before passing through the last linear layer, the output vectors that are generated by each head are combined into a single vector by means of concatenation. In principle, each head would learn something unique, which would result in the encoder model having a greater capacity for representation.

Multi-headed attention is just self-attention applied multiple times in different ways. The goal is to capture the different contextual representations via these different heads. By using multi-headed attention, the representations we get for a specific word are very rich.

The Residual Connections, Layer Normalization, and Feed-Forward Network

The original positional input embedding is then given the multi-headed attention output vector as an additional component. This type of connection is known as a residual connection. You can think of this step as addition of input (positional encoding in this case) to the output (multi-headed attention output in this case) . After this an operation known as layer normalization is performed on the output of the residual connection. The goal of layer normalization is for improving the performance of training.

Post normalization, the output is passed via a feed-forward network, and then the result of this feed-forward network is normalized with input as the data fed to the feed-forward network.

Because they enable gradients to move directly across the network, the residual connections are beneficial to the network's training process. The layer normalizations are what are responsible for stabilizing the network, which ultimately leads to a significant cut in the amount of training time required.

This is more or less at a high level how the encoder works. Next, we discuss in brief the decoder component of transformers.

Decoder

It is the responsibility of the decoder to produce text sequences. The decoder is similar to encoders in having the layers like

1. Multi-headed attention layer

2. Add and norm layers

3. Feed-forward layer

In addition it has a linear layer with a softmax classifier to emit probabilities of an output. This is where the generative part comes into play.

The decoder works by taking the starting tokenized word and then previous outputs if any and combining it with the output of the encoder.

The different aspects involved in the decoding process are explained in the following.

Input Embeddings and Positional Encoding

To a large extent, the beginning of the decoder is identical to the beginning of the encoder. In order to obtain positional embeddings, the input is first placed via an embedding layer and then a positional encoding layer. The positional embeddings are then sent through to the first multi-head attention layer.

First Layer of Multi-headed Attention

This layer, though similar in name to what we used in encoders, is a bit different in functionality. The reason is that the decoder only has access to words that come prior to the current word in the sentence. It's not supposed to see what word comes next in the sequence.

There needs to be a way for us to avoid computing attention scores for terms in the future. The technique in question is known as masking. By using a lookahead mask, you can restrict the decoder from looking at tokens that are yet to come. The mask is included both before and after the softmax calculation, which takes place after the scores have been scaled. We won't discuss the mathematical details of the lookahead mask in this book. The basic idea of the mask is to only calculate the attention score for the current word based on previous words and not for future words in the sentence.

Second Layer of Multi-headed Attention

This layer takes the output from the first multi-headed attention layer of the decoder and combines this with the output of the encoder. This will enable the decoder to understand better as to which components of encoder output to attend to. The output of this multi-headed attention layer is passed via a feed-forward network.

Linear Classifier and Final Softmax for Output Probabilities

The output of the previous multi-headed attention layer and feed-forward network is again normalized and passed to a linear layer with a softmax component for emitting the probabilities – as an example, the probability of what could be the next word in a specific sequence of words. This is where the generative aspect of the architecture shines.

Summary

This chapter explained how transformers work. Also it detailed out the power of the attention mechanism that is harnessed by transformers so that they can make more accurate predictions. Similar goals can be pursued by recurrent neural networks, but their limited short-term memory makes progress toward these goals difficult. If you wish to encode a long sequence or produce a long sequence, transformers may be a better option for you. The natural language processing sector is able to accomplish results that have never been seen before as a direct result of the transformer design.

In the next chapter, we look into how transformers work from the point of view of code in more detail. We introduce the huggingface ecosystem, which is one of the major open source repositories for transformer models.

CHAPTER 3

BERT

In this chapter, you will learn one of the implementations of the transformer architecture, developed by Google, called BERT.

Recent work done by researchers at Google AI Language resulted in the publication of a paper known as "BERT (Bidirectional Encoder Representations from Transformers)."

The most important technical advancement that BERT has made is the application of bidirectional training of the popular attention model, transformer, to language modeling. According to the findings of the study on language models, a language model that is trained in both directions at once is able to have a greater awareness of the flow and context of language than models trained in just one direction. The researchers describe a unique training method that they call masked language modeling (MLM) in the publication. This method enables bidirectional training in models, which was previously difficult to do.

BERT also allows us to do transfer learning, which allows the pretrained BERT model to be used in a variety of natural language applications.

Depending on the scale of the model architecture, there are two different pretrained versions of BERT, which are as follows.

BERT-Base has a total of 110 million parameters, 12 attention heads, 768 hidden nodes, and 12 layers.

BERT-Large is characterized by having 24 layers, 1024 hidden nodes, 16 attention heads, and 340 million parameter values.

© Shashank Mohan Jain 2022
S. M. Jain, *Introduction to Transformers for NLP*,
https://doi.org/10.1007/978-1-4842-8844-3_3

Workings of BERT

BERT is based on the transformer architecture, which internally makes use of the attention mechanism as discussed in Chapter 2 . BERT's beauty lies in understanding the context in a sentence and representing a word, keeping the context into account. This means that a word like *bank* when used in the context of finance has a different representation than when the same word *bank* is used as in a river bank. BERT only uses the encoder mechanism of the transformer architecture and is mainly used to create better word representations, which can then be used for more downstream applications. The detailed working of transformers was covered in Chapter 2.

The input to the transformer encoder is a sentence of words, and the architecture of BERT is such that it can read in both directions to capture context in words, meaning words that occur both prior to the word in focus and later in the sentence. This overcomes sequential reading, which happens in, say, LSTM-type architectures (although bidirectional LSTMs are possible, they are complex and still have sequential processing) .

In the process of training language models, one of the challenges that must be overcome is identifying a prediction goal. A large number of models provide predictions about the following word in a sequence (e.g., "The sun is sinking in the ___"), which is a directional approach that limits context learning by its very nature. BERT employs two different training tactics in order to overcome this obstacle.

Masked LM (MLM)

Providing BERT with a sentence and then optimizing the weights included inside it in order to produce the same sentence on the other side is what constitutes MLM. Therefore, we give BERT a sentence and ask it to produce the same sentence as the input. Before we really provide BERT that input sentence, we cover up a few tokens.

Before being fed into BERT, word sequences have 15% of their words substituted with [MASK] tokens. This is done before the word sequences are fed into BERT. The model will then make an attempt to make a prediction of the possible value of the masked words. It uses the context provided by surrounding words, both in the left and right directions.

To predict the masked sentence, we will need an additional layer on top of the encoder, which would help in classifying the masked sentence.

The loss function used takes into consideration only the prediction of the masked values. As a direct result of this, the model converges more slowly than directed models do; however, the higher context awareness that it possesses makes up for this shortcoming.

Figure 3-1 shows how training a masked language model works. Around 15% of the input tokens are masked, and the feed-forward neural network then is trained to predict only these masked tokens.

Figure 3-1. Masked language modeling

An example of how training a masked language model works, in which only a few masked tokens are predicted, not each input masked word

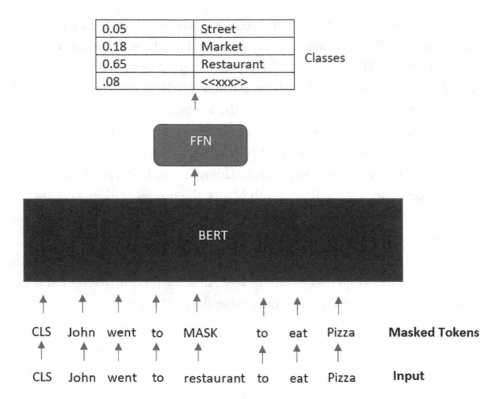

0.05	Street
0.18	Market
0.65	Restaurant
.08	<<xxx>>

Figure 3-1. *Masked language model*

An example of how inference in MLM works is shown in Figure 3-2, which shows how a masked language model can be used to predict the masked word.

```
from transformers import pipeline
unmasker = pipeline('fill-mask', model='bert-base-cased')
unmasker("I am going to [MASK] to eat pizza.")
```

```
Some weights of the model checkpoint at bert-base-cased were
- This IS expected if you are initializing BertForMaskedLM f
- This IS NOT expected if you are initializing BertForMaskec
[{'score': 0.8603612184524536,
 'sequence': 'I am going to have to eat pizza.',
 'token': 1138,
 'token_str': 'have'},
 {'score': 0.032728005200624466,
 'sequence': 'I am going to need to eat pizza.',
 'token': 1444,
 'token_str': 'need'},
 {'score': 0.027744023129343987,
 'sequence': 'I am going to get to eat pizza.',
 'token': 1243,
 'token_str': 'get'},
 {'score': 0.017428388819098473,
 'sequence': 'I am going to want to eat pizza.',
 'token': 1328,
 'token_str': 'want'},
```

Figure 3-2. *Masked language model usage for token prediction*

This is a screenshot from using the huggingface library, which we will cover in the next chapter.

Next Sentence Prediction

The Next Sentence Prediction (NSP) protocol entails providing BERT with two sentences, designated as sentence A and sentence B. Then, we inquire of BERT, "Hey, does sentence B come after sentence A?" – and depending on the situation, BERT will either say IsNextSentence or NotNextSentence.

41

Let us consider three sentences in our dataset:

1. John went to the restaurant.

2. The kite is flying high in the sky.

3. John ordered a pizza.

Now when we look at these three sentences, we can easily make out that sentence 2 does not follow sentence 1 and, on the contrary, sentence 3 follows sentence 1. This mode of reasoning, which involves dependencies between phrases that span extended time periods, is taught to BERT via NSP.

For training with NSP in mind, the BERT architecture takes positive pairs and negative pairs as inputs.

The positive pair constitutes sentences that are related to each other, whereas the negative pair constitutes sentences not related each other in the sequence. These negative and positive samples constitute 50% each in the dataset.

Before the input is added to the model, it is processed in the following manner so that it can better assist the model in differentiating between the two sentences used in training:

1. At the beginning of the very first sentence, a [CLS] token is inserted, and at the very end of each sentence, a [SEP] token is placed.

2. Each token now has a sentence embedding that indicates whether it belongs to sentence A or sentence B. Token embeddings and sentence embeddings are conceptually comparable representations of a vocabulary size of 2.

3. Each token receives an additional embedding called a positional embedding so that its position in the sequence may be determined. This is explained in Chapter 2, in the section "Positional Encoding."

Inference in NSP

To demonstrate inference via NSP, we take the three sentences:

1. John went to the restaurant.

2. The kite is flying high in the sky.

3. John ordered a pizza.

Calculate the probability of whether sentence 2 follows sentence 1 and the probability of sentence 3 following sentence 1.

We will again use the huggingface library (which we will cover in the next chapter) to calculate the individual probabilities.

See sample code to be used in Google Colab. (Don't worry about code as of now as there will be plenty of examples in Chapter 5.)

Listing 3-1. BERT for sentence prediction

```
from torch.nn.functional import softmax
from transformers import BertForNextSentencePrediction,
BertTokenizer

mdl = BertForNextSentencePrediction.from_pretrained('bert-
base-cased')
brt_tkn = BertTokenizer.from_pretrained('bert-base-cased')

sentenceA = 'John went to the restaurant'

sentenceB = 'The kite is flying high in the sky'

encoded = brt_tkn.encode_plus(sentenceA, text_pair=sentenceB,
return_tensors='pt')

sentence_relationship_logits = mdl(**encoded)[0]

probablities = softmax(sentence_relationship_logits, dim=1)

print(probablities)
```

We get output as

```
tensor([[0.0926, 0.9074]], grad_fn=<SoftmaxBackward0>)
```

This shows that there is a very low probability that sentence 2 follows sentence 1.

Now we run the same inference for sentence 1 and sentence 3:

```
sentenceA = 'John went to the restaurant'
sentenceB = 'John ordered a pizza'
```

We get output as

```
tensor([[9.9998e-01, 2.2391e-05]], grad_fn=<SoftmaxBackward0>)
```

This shows that there is a high probability of sentence 3 following sentence 1.

BERT Pretrained Models

BERT is based on the transformer architecture and uses the encoder mechanism primarily. It has many variations including the following.

BERT-Base has a smaller size for its transformer blocks and its hidden layers than OpenAI GPT does, but it has the same overall model size (12 transformer blocks, 12 attention heads, and a size of 768 for the hidden layer).

BERT-Large is an enormous network that accomplishes state-of-the-art results on NLP tasks. It has twice as many attention layers as BERT-Base (24 transformer blocks, 16 attention heads, and a size of 1024 for the hidden layer).

The pretrained BERT models are available at huggingface and can be used directly for the fine-tuning of downstream tasks.

A few of the pretrained BERT models available on huggingface are shown in Figure 3-3.

bert-base-uncased	12-layer, 768-hidden, 12-heads, 110M parameters. Trained on lower-cased English text.
bert-large-uncased	24-layer, 1024-hidden, 16-heads, 340M parameters. Trained on lower-cased English text.
bert-base-cased	12-layer, 768-hidden, 12-heads, 110M parameters. Trained on cased English text.
bert-large-cased	24-layer, 1024-hidden, 16-heads, 340M parameters. Trained on cased English text.
bert-base-multilingual-uncased	(Original, not recommended) 12-layer, 768-hidden, 12-heads, 110M parameters. Trained on lower-cased text in the top 102 languages with the largest Wikipedias (see details).
bert-base-multilingual-cased	(New, recommended) 12-layer, 768-hidden, 12-heads, 110M parameters. Trained on cased text in the top 104 languages with the largest Wikipedias (see details).

Figure 3-3. *Bert models from huggingface*

BERT Input Representations

- Always consider the first token in a sequence to be a special classification token (also abbreviated CLS). For the purposes of the classification task, the final hidden state that corresponds to this token will be used.

- The [SEP] token serves to demarcate the break between the two sentences.

- Whenever a sentence pair is being processed, an additional segment embedding is going to be added. This embedding will indicate whether the token belongs to sentence A or sentence B.

- The input representation of a given token is constructed by adding positional embedding to the representation of the token.

Use Cases for BERT

BERT can be used for a variety of downstream tasks once we have the proper representations generated by the encoder. These tasks include sentiment analysis, summarization, Q&A, text-to-SQL generation, etc.

Apart from using the pretrained BERT models, we can also fine-tune them with our specific text datasets. This will allow us to take advantage of our own data.

In contrast with other large learning models, such as GPT3, the source code for BERT is freely available to the public and can be viewed on GitHub. This makes it possible for BERT to be utilized in a wider variety of contexts all over the world. This completely changes the dynamic!

It is now possible for developers to quickly get a cutting-edge model like BERT up and running without having to invest a significant amount of time or money in the process.

Also, developers can focus on fine-tuning BERT to tailor the model's performance to the specific requirements of their individual projects.

If one doesn't want to spend time fine-tuning BERT, they should be aware that there are thousands of open source and free BERT models that have already been pretrained and are currently available for specific use cases.

BERT models have been pretrained for a variety of tasks like

1. Analysis of user sentiment on Twitter and other social media

2. Toxic comment detection

3. Speech-to-text

4. Question and answering

And many more.

We take a tweet classification example of BERT in the following using the huggingface library.

Sentiment Analysis on Tweets

The BERTweet language model is the first publicly available large-scale model that has been pretrained for English tweets. The pretraining technique for RoBERTa is utilized to instruct BERTweet's training. The pretraining dataset for BERTweet comprises 850 million English tweets (16 billion word tokens and 80 gigabytes), 845 million tweets streamed from January 1, 2012, to August 8, 2019, as well as 5 million tweets relating to the COVID-19 pandemic.

Listing 3-2. Sentiment analysis using BERT

```
model = pipeline('sentiment-analysis', model="finiteautomata/
bertweet-base-sentiment-analysis")
```

By running this code, we get the following output as shown in Figure 3-4 .

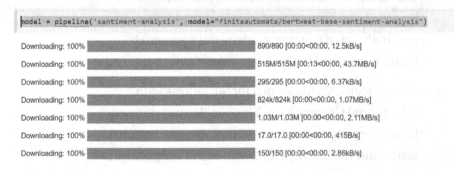

Figure 3-4. *Execution of Listing 3-2. This shows downloading of models and tokenizers*

We then feed two tweets into the model:

```
data = ["idk about you guys but i'm having more fun during the
bear than I was having in the bull.","At least in the bear
market it's down only. Bull market is up and down"]
model(data)
```

This gives the following output:

```
[{'label': 'POS', 'score': 0.9888675808906555}, {'label':
'NEG', 'score': 0.7523579001426697}]
```

This shows how the Bertweet model is able to classify the preceding two tweets.

The POS score shows the positive sentiment, and the NEG score shows the negative sentiment.

Performance of BERT on a Variety of Common Language Tasks

BERT performance was measured on the following few benchmarks:

1. SQuAD (Stanford Question Answering Dataset): This is a question and answering dataset from Stanford. Performance of BERT on this dataset was much ahead of existing models and also much better compared with a human.

2. SWAG: Here, SWAG stands for Situations with Adversarial Generations. This is a dataset derived from everyday scenarios and mainly tests common sense and reasoning. Here, also BERT outperformed other models as well as human performance.

3. GLUE: Stands for General Language Understanding
   Evaluation. This evaluates how a model works
   in terms of understanding a specific language.
   There are some specific tasks on which a model is
   evaluated here. Here, also the performance of BERT
   was exceptionally good.

Summary

BERT is an extremely complicated and cutting-edge language model that assists individuals in automating their language comprehension. It is supported by training on massive amounts of data and leveraging the transformer architecture to revolutionize the field of natural language processing, which enables it to accomplish state-of-the-art performance.

It appears that untouched NLP milestones have a bright future, thanks to the open source library that BERT provides, as well as the efforts that the incredible AI community is making to continue to improve and share new BERT models.

CHAPTER 4

Hugging Face

If you have even a passing familiarity with the advancements that have been made in the fields of machine learning and artificial intelligence in the years since 2018, you have almost certainly become aware of the tremendous strides that have been taken in the field of natural language processing (also known as NLP). Most of the progress in this area can be attributed to large language models, also known as LLMs. The architecture behind these LLMs is the transformer's encoder-decoder, which we discussed in Chapter 2.

Success of transformers came from the architecture's ability to process input data in parallel as well as having a better contextual understanding via the attention mechanism. We already referred to Vaswani's "Attention Is All You Need" paper in previous chapters. Before the emergence of transformers, the context was captured by vanilla RNN or LSTM without attention.

Hugging Face is now widely recognized as a one-stop shop for all things related to natural language processing (NLP), offering not only datasets and pretrained models but also a community and even a course.

Hugging Face is a new company that has been built on principles of using open source software and data. In true sense, the revolution in NLP started with democratization of the NLP models based on the transformer architecture. Hugging Face turned out to be a pioneer in not just open sourcing these models but also providing a handy and easy-to-use abstraction in the form of the Transformers library, which made it really easy to consume and infer these models.

© Shashank Mohan Jain 2022
S. M. Jain, *Introduction to Transformers for NLP*,
https://doi.org/10.1007/978-1-4842-8844-3_4

Hugging Face provided a central place or hub for model developers to publish the models in the huggingface repository, which can then be consumed by consumers who are looking to build applications on top of these models. As an example, the BERT (Bidirectional Encoder Representations from Transformers) model was contributed by Google to huggingface. This then allowed a community of users to consume these models in their applications. Then came GPT models from OpenAI like GPT2, which are generative models and allowed the end user to write applications that can generate, say, stories, novels, etc. GPT2 is also a part of the huggingface ecosystem. Hugging Face provided not only APIs for consuming those models but also a way to fine-tune them with our own dataset and monitor and benchmark these models. So in a nutshell, the emergence of an ecosystem like huggingface has really opened up a plethora of opportunities for developers intending to build applications on top of natural language processing–based models.

Currently the scope of these models is not just limited to text processing. We now see the emergence of vision transformers and transformer-based models for audio. People are building applications for music generation or voice cloning in the audio domain and using them for fake image generation in the case of vision use cases. There are also models that have mined scientific literature and can be used for extracting knowledge from scientific journals. Similarly, models based on law-related documents have emerged, and people can use them to build a question and answering system on top of them. These models come in a variety of sizes based on the underlying architectures being used. As an example, the GPT3 model has around 175 billion parameters.

Similarly, the size and scope of their training datasets have both increased significantly. For instance, the original transformer was replaced by the much larger Transformer-XL; the number of parameters in BERT-Base increased from 110 million to 340 million in Bert-Large; and the GPT2 model, which had 1.5 billion parameters, was replaced by the GPT3 model that has 175 billion parameters. China launched a model named

Wu Dao 2.0, which has around 1.75 trillion parameters. The proponents of scaling opine that as we increase the sizes of these models, we will also be reaching our goal of artificial general intelligence (AGI).

To give an example of infrastructure needed for such large models, GPT3 was trained on a super-computer with more than 10000 GPUs. This means training these models only lies in the realm of big companies. Now with the ability to consume most of these models, the end user becomes part of the application development ecosystem based around these large language models.

Features of the Hugging Face Platform

Because the Hugging Face platform is predicated on the idea of attention-based transformer models, it should come as no surprise that the Transformers library is at the center of the Hugging Face ecosystem. The accompanying Datasets and Tokenizers libraries offer assistance to the Transformers library. Keep in mind that transformers are unable to comprehend text in its original form, which is a string of characters. Since our inputs to transformers are in text, this text has to be encoded in a way that makes it consumable via the transformer-based neural network architecture. For this we make use of huggingface-provided APIs for tokenizing, which are known as tokenizers.

Apart from tokenizing, we might need to use some custom datasets to either fine-tune existing models or train the models from scratch. To have a uniformity in architecture, huggingface provides an abstraction for datasets via the Datasets API. The user can then have their own datasets, upload the datasets, train/fine-tune the models, and also upload the trained models all just by using the huggingface APIs. This is nothing less than revolutionary.

Components of Hugging Face

The huggingface library is based on a set of rich abstractions, which abstract out the complexity of creating applications based on natural language processing. These abstractions allow us a single interface to load models, use tokenizers, use datasets across a variety of models, tokenizers and datasets. This is really an empowering experience for the developer whose job is simplified by using these abstractions. We discuss some of these abstractions in the following subsections.

Pipelines

Pipelines provide a powerful and convenient abstraction for consuming the pretrained models from the huggingface model repository. It offers a straightforward application programming interface (API) that is dedicated to a variety of tasks:

1. Determine whether the overall sentiment of the sentence can be characterized as positive or negative.

2. Question and answering takes a question and pulls an answer out of the text that corresponds to it.

3. The masked language modeling technique suggests possible words to fill masked input with the given context.

4. The named entity recognition program will automatically assign a label to each of the tokens that are included in the input.

5. Reducing a longer piece of writing or an article into a more concise summary is referred to as summarization.

Pipelines are an abstraction built on top of three huggingface components, namely

1. Tokenizer

2. Model

3. Post-processor

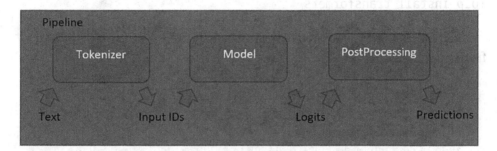

Figure 4-1. *Workflow of a huggingface pipeline*

Tokenizer

The first component in the pipeline is a tokenizer, which takes raw text as input and converts it into numbers so that the model can interpret them.

The tokenizer is responsible for

1. The process of separating the input into individual tokens, which may be words, sub-words, or symbols (such as punctuations)

2. Converting each token into an integer

3. Introducing new variables into the model that might prove to be of some use

When the model was trained, we had the need to tokenize the inputs. At that time there would have been a use of a certain tokenizer. We need to make sure that during usage of this model on actual inputs, we use the same tokenizer. This task is made easy by the AutoTokenizer class, which will automatically load the tokenizer used during the training. This makes the life simple for a developer considerably.

We load the tokenizer used to pretrain the GPT2 model in the following code.

First, install the Transformers library in Google Colab:

```
!pip install transformers
```

Figure 4-2 shows the installation process of the Transformers library from huggingface.

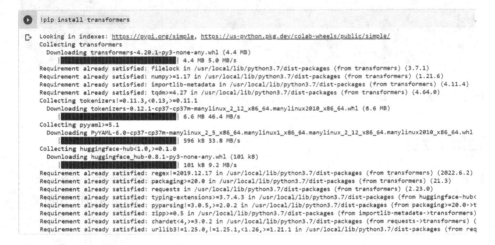

Figure 4-2. *Installation of the Transformers library in Google Colab*

Next, add this code.

Listing 4-1. Code for a simple tokenizer

```
from transformers import AutoTokenizer

tokenizer = AutoTokenizer.from_pretrained("gpt2")

encoding = tokenizer("This is my first stab at AutoTokenizer")
print(encoding)
```

Executing Listing 4-1 in Google Colab results in the output shown in Figure 4-3.

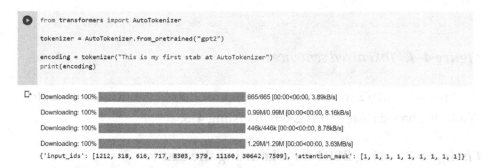

Figure 4-3. *Downloading of a tokenizer*

The tokenizer will provide a dictionary with the following entries:

The numerical representations of your tokens are referred to as input IDs.

The attention mask is a mask that specifies which tokens need to have attention paid to them.

We can also pass multiple strings as inputs to the tokenizer. An example is shown in Listing 4-2, which tokenizes the text in two sentences.

Listing 4-2. Code for using a tokenizer for tokenizing text

```
from transformers import AutoTokenizer

tokenizer = AutoTokenizer.from_pretrained("gpt2")

encoding = tokenizer("This is my first stab at
AutoTokenizer","life is what happens when you are planning
other things")
print(encoding)
```

This would give tokens for individual sentences as shown in Figure 4-4.

```
from transformers import AutoTokenizer

tokenizer = AutoTokenizer.from_pretrained("gpt2")

encoding = tokenizer("This is my first stab at AutoTokenizer","life is what happens when you are planning other things")
print(encoding)


{'input_ids': [1212, 318, 616, 717, 8303, 379, 11160, 30642, 7509, 6042, 318, 644, 4325, 618, 345, 389, 5410, 584, 1243], 'attention_mask': [1, 1,
```

Figure 4-4. *Tokenized sentences*

Instead of GPT2 we can also use other models like BERT. An example
of a BERT-based tokenizer is given in Listing 4-3.

Listing 4-3. Using BERT for tokenizing the text

```
from transformers import AutoTokenizer

tokenizer = AutoTokenizer.from_pretrained("bert-base-cased")

encoding = tokenizer("This is my first stab at AutoTokenizer")
print(encoding)
```

Figure 4-5 shows the output of a BERT-based tokenizer. This is
achieved by running Listing 4-3 in Google Colab.

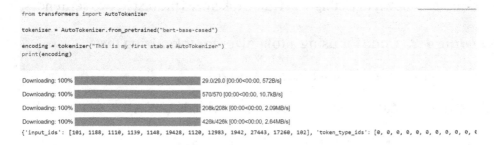

Figure 4-5. *Execution of a BERT-based tokenizer*

The output encoding is

```
{'input_ids': [101, 1188, 1110, 1139, 1148, 19428, 1120, 12983,
1942, 27443, 17260, 102], 'token_type_ids': [0, 0, 0, 0, 0, 0,
0, 0, 0, 0, 0, 0], 'attention_mask': [1, 1, 1, 1, 1, 1, 1, 1,
1, 1, 1, 1]}
```

This returns a dictionary that contains the following three significant items:

1. The indices that correspond to each token in the sentence are denoted by the input ids variable.

2. The value of the attention mask specifies whether or not a token needs to be attended to.

3. When there is more than one sequence, the token type ids variable is used to determine which sequence a token is a part of.

We can get back the input by decoding the input_ids as shown in the following:

```
tokenizer.decode(encoding["input_ids"])
```

Its output is as shown in Figure 4-6.

 `tokenizer.decode(encoding["input_ids"])`

`'[CLS] This is my first stab at AutoTokenizer [SEP]'`

Figure 4-6. *Shows the decoding process by taking tokens as inputs and returning the text as output*

As can be seen, the tokenizer inserted two specialized tokens into the sentence. These tokens are known as CLS and SEP, which stand for classifier and separator, respectively. The tokenizer will take care of adding any necessary special tokens for you, provided that the model in question actually requires them.

Let's pass multiple sentences to this tokenizer as shown in Listing 4-4.

Listing 4-4. This code takes two sentences as input and generates tokens for them

```
from transformers import AutoTokenizer

tokenizer = AutoTokenizer.from_pretrained("bert-base-cased")

encoding = tokenizer("This is my first stab at
AutoTokenizer","life is what happens when you are planning
other things")
print(encoding)
```

This results in multiple encodings:

```
{'input_ids': [101, 1188, 1110, 1139, 1148, 19428, 1120, 12983,
1942, 27443, 17260, 102, 1297, 1110, 1184, 5940, 1165, 1128,
1132, 3693, 1168, 1614, 102], 'token_type_ids': [0, 0, 0, 0,
0, 0, 0, 0, 0, 0, 0, 0, 1, 1, 1, 1, 1, 1, 1, 1, 1, 1, 1],
'attention_mask': [1, 1, 1, 1, 1, 1, 1, 1, 1, 1, 1, 1, 1, 1, 1,
1, 1, 1, 1, 1, 1, 1, 1]}
```

Padding

When we process a group of sentences, their individual lengths do not always remain consistent. The inputs to the models need to have the same size as this is based on the underlying standard architecture. This presents a problem. The addition of a padding token to a sentence that contains an insufficient number of tokens is an example of the strategy known as "padding."

Now we run the following example with multiple inputs with padding set to true as shown in Listing 4-5.

Listing 4-5. This code shows how padding for tokenizers works

```
from transformers import AutoTokenizer

bert_tk = AutoTokenizer.from_pretrained("bert-base-cased")
sentences=["This is my first stab at AutoTokenizer","life is
what happens when you are planning other things","how are you"]

encoding = bert_tk(sentences,padding=True)
print(encoding)
```

This gives the output as in the following:

```
{'input_ids': [[101, 1188, 1110, 1139, 1148, 19428, 1120,
12983, 1942, 27443, 17260, 102], [101, 1297, 1110, 1184, 5940,
1165, 1128, 1132, 3693, 1168, 1614, 102], [101, 1293, 1132,
1128, 102, 0, 0, 0, 0, 0, 0, 0]], 'token_type_ids': [[0, 0, 0,
0, 0, 0, 0, 0, 0, 0, 0, 0], [0, 0, 0, 0, 0, 0, 0, 0, 0, 0, 0,
0], [0, 0, 0, 0, 0, 0, 0, 0, 0, 0, 0, 0]], 'attention_mask':
[[1, 1, 1, 1, 1, 1, 1, 1, 1, 1, 1, 1], [1, 1, 1, 1, 1, 1, 1, 1,
1, 1, 1, 1], [1, 1, 1, 1, 1, 0, 0, 0, 0, 0, 0, 0]]}
```

As we can see, the third sentence was shorter in length, and thereby the tokenizer padded it with zeroes.

Truncation

It's possible that a model can't deal with a sequence that's too long sometimes. In this particular scenario, you will be required to condense the sequence down to a more manageable length.

If you want to truncate a sequence so that it is the maximum length that the model will accept, set the truncation parameter to true as shown in Listing 4-6.

Listing 4-6. This code shows how the truncation flag works in tokenizers

```
from transformers import AutoTokenizer

bert_base_tokenizer = AutoTokenizer.from_pretrained("bert-
base-cased")
sentences=["This is my first stab at AutoTokenizer","life is
what happens when you are planning other things. so plan life
accordingly","how are you"]

encoding = bert_base_tokenizer(sentences,padding=True,trunca
tion=True)
print(encoding)
```

The output is shown in the following:

```
{'input_ids': [[101, 1188, 1110, 1139, 1148, 19428, 1120,
12983, 1942, 27443, 17260, 102, 0, 0, 0, 0, 0], [101, 1297,
1110, 1184, 5940, 1165, 1128, 1132, 3693, 1168, 1614, 119,
1177, 2197, 1297, 17472, 102], [101, 1293, 1132, 1128, 102, 0,
0, 0, 0, 0, 0, 0, 0, 0, 0, 0, 0]], 'token_type_ids': [[0, 0, 0,
0, 0, 0, 0, 0, 0, 0, 0, 0, 0, 0, 0, 0, 0], [0, 0, 0, 0, 0, 0,
0, 0, 0, 0, 0, 0, 0, 0, 0, 0, 0], [0, 0, 0, 0, 0, 0, 0, 0, 0,
0, 0, 0, 0, 0, 0, 0, 0]], 'attention_mask': [[1, 1, 1, 1, 1, 1,
1, 1, 1, 1, 1, 1, 0, 0, 0, 0, 0], [1, 1, 1, 1, 1, 1, 1, 1, 1,
1, 1, 1, 1, 1, 1, 1, 1], [1, 1, 1, 1, 1, 0, 0, 0, 0, 0, 0, 0,
0, 0, 0, 0, 0]]}
```

The next stage in the pipeline is the model.

AutoModel

We will check how AutoModel makes life simpler for us in terms of loading pretrained models.

The process of loading pretrained instances is made straightforward and unified by the Transformers library. This indicates that you are able to load an AutoModel in the same way that you load an AutoTokenizer. The sole distinction lies in selecting the appropriate AutoModel for the task at hand.

If we take an example of text classification, the way we load the model is shown in the following.

We will follow these steps:

1. Create an instance of a tokenizer and a model based on the name of the checkpoint. The model is determined to be a BERT model, and then weights that have been saved in the checkpoint are loaded into it.

2. Get the tokens and pass it to the model.

3. The model returns the logits.

4. Apply a softmax to calculate the probability of the class in which the sentence is classified (negative or positive for our following example).

Listing 4-7 is broken into multiple sections wherein Listing 4-7-1 explains how a tokenizer is loaded via the Transformers library.

Listing 4-7-1. Loading the tokenizer

```
from transformers import AutoTokenizer

tokenizer = AutoTokenizer.from_pretrained("siebert/sentiment-
roberta-large-english")
sentences=["This is my first stab at AutoTokenizer","life is
what happens when you are planning other things. so plan life
accordingly","this is not tasty at all"]
```

```
encoding = tokenizer(sentences,padding=True,truncation=True,ret
urn_tensors="pt")
print(encoding)
```

This gives the following output:

```
{'input_ids': tensor([[    0,   713,    16,   127,    78,
16735,    23,  8229, 45643,  6315,
          2,     1,     1,     1,     1,     1,     1],
      [    0,  5367,    16,    99,  2594,    77,    47,
32,  1884,    97,
        383,     4,    98,   563,   301, 14649,     2],
      [    0,  9226,    16,    45, 22307,    23,  1250,
2,     1,     1,
          1,     1,     1,     1,     1,     1,     1]]),
        'attention_mask': tensor([[1, 1, 1, 1, 1, 1, 1,
        1, 1, 1, 1, 0, 0, 0, 0, 0, 0],
      [1, 1, 1, 1, 1, 1, 1, 1, 1, 1, 1, 1, 1, 1, 1, 1, 1],
      [1, 1, 1, 1, 1, 1, 1, 1, 0, 0, 0, 0, 0, 0, 0, 0, 0]])}
```

Listing 4-7-2 shows how a model is loaded via the Transformers library.

for loading the model

Listing 4-7-2. Loading the model

```
from transformers import AutoModelForSequenceClassification

model_name = "siebert/sentiment-roberta-large-english"
pt_model = AutoModelForSequenceClassification.from_
pretrained(model_name)
```

Listing 4-7-3. Generate the logits

```
#For printing the encoding

pt_outputs = pt_model(**encoding)
print (pt_outputs)

SequenceClassifierOutput(loss=None, logits=tensor([[ 3.0351,
-2.1955], [-3.6225,   2.7819], [ 3.9581, -3.6334]]), grad_
fn=<AddmmBackward0>), hidden_states=None, attentions=None)
```

   #For printing the logits

```
logits=pt_outputs.logits
print (logits)
```

   This outputs

```
tensor([[ 3.0351, -2.1955],
        [-3.6225,  2.7819],
        [ 3.9581, -3.6334]], grad_fn=<AddmmBackward0>)
```

Finally, we do a softmax to print the output probabilities as illustrated in Listing 4-7-4.

Listing 4-7-4. Print probabilities specific to the individual class of sentiment

```
output = torch.softmax(logits, dim=1).tolist()[1]
print(output)
```

   We get the following output:

```
[[0.9946781396865845, 0.005321894306689501],
[0.001651538535952568, 0.9983484745025635],
[0.9994955062866211, 0.0005045001162216067]]
```

We can see the output probabilities for all three sentences:

```
This is my first stab at AutoTokenizer
```

The first column gives the probability of sentiment being negative and the second column of sentiment being positive:

```
Score [0.9946781396865845, 0.005321894306689501]
```

This reflects a negative sentiment in the sentence.

Next is for the second sentence:

```
life is what happens when you are planning other things. so
plan life accordingly
[0.001651538535952568, 0.9983484745025635]
```

This reflects a positive sentiment in the sentence.

Now we look into a wrapper class known as pipeline, which can be used to achieve the same task with less code as shown in Listing 4-8.

Listing 4-8. This code shows how to use the pipeline API for doing sentiment analysis

```
from transformers import pipeline
# create a pipeline instance with a tokenizer and model
roberta_pipe = pipeline(
    "sentiment-analysis",
    model="siebert/sentiment-roberta-large-english",
    tokenizer="siebert/sentiment-roberta-large-english",
    return_all_scores = True
)

# analyze the sentiment for the 3 sentences we used in the
preceding example
roberta_pipe(sentences)
```

We get the following output:

```
[[{'label': 'NEGATIVE', 'score': 0.9946781396865845}, {'label':
'POSITIVE', 'score': 0.005321894306689501}],
[{'label': 'NEGATIVE', 'score': 0.001651539234444499},
{'label': 'POSITIVE', 'score': 0.9983484745025635}],
[{'label': 'NEGATIVE', 'score': 0.9994955062866211}, {'label':
'POSITIVE', 'score': 0.0005045001744292676}]]
```

We can check that the outputs match when we didn't use and used the pipeline class in our code.

Summary

In this chapter we discussed the architecture of the huggingface library and its components like tokenizers and models. We also learned how we can use these components to do a simple task like analyzing the sentiments of the sentences. In the next chapter, we will take more examples of doing different kinds of tasks using the Transformers library.

CHAPTER 5

Tasks Using the Hugging Face Library

So far we have seen how transformers can be used with the huggingface Transformers library at a very elementary level. We will now start to see how we can use the library for different tasks related to not just text but also audio and images.

But before we move on with this, we will introduce you to Gradio, which is a library for building UI on top of huggingface.

Gradio: An Introduction

Gradio is a web framework specially built for deploying and inferencing machine learning models. Gradio allows us to have our ML models exposed quickly over a web interface without a requirement of learning too much coding. With acquisition of Gradio, Hugging Face has moved one step ahead in providing the huggingface community an easy interface to deploy and provide UI over the huggingface models.

© Shashank Mohan Jain 2022
S. M. Jain, *Introduction to Transformers for NLP*,
https://doi.org/10.1007/978-1-4842-8844-3_5

In this chapter we will make use of huggingface spaces, which provide us an interface to quickly deploy and provide our application (built using the huggingface APIs), a web front end that can then be used by end users to interact with our application.

Creating a Space on Hugging Face

To create a space on the huggingface infra, we need to have an account with huggingface. This can be done by navigating to `https://huggingface.co/` and creating an account there. Once the account is created, we can click the colored circle on the extreme right as shown in Figure 5-1.

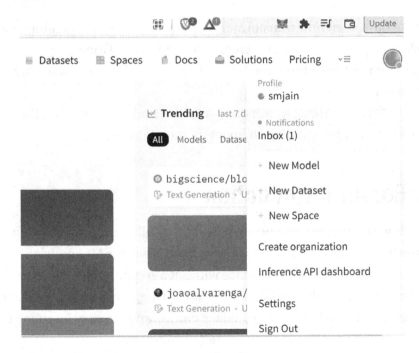

Figure 5-1. Hugging Face screen after login

Click New space, and we see a screen as shown in Figure 5-2.

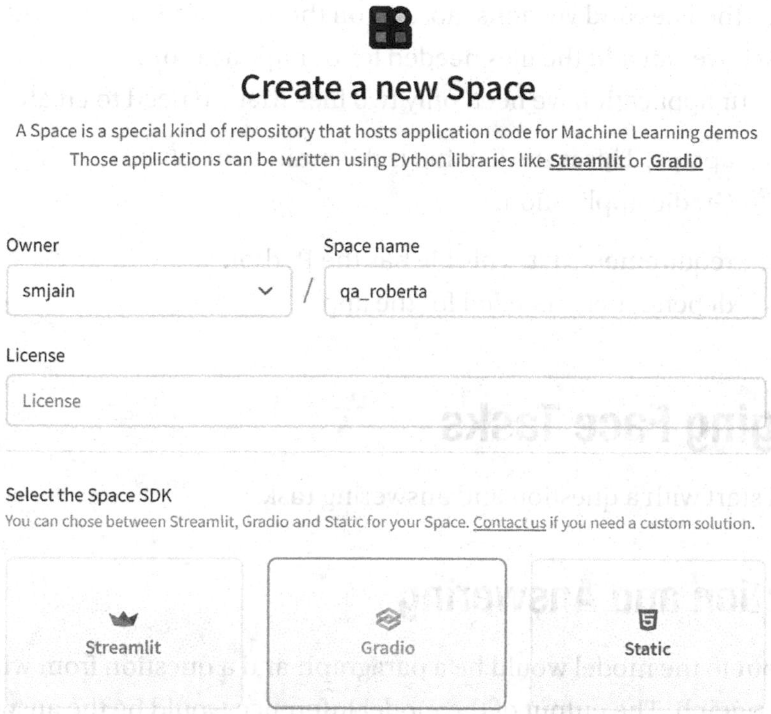

Create a new Space

A Space is a special kind of repository that hosts application code for Machine Learning demos
Those applications can be written using Python libraries like **Streamlit** or **Gradio**

Owner   Space name

| smjain   ∨ | / | qa_roberta |

License

| License |

Select the Space SDK
You can chose between Streamlit, Gradio and Static for your Space. Contact us if you need a custom solution.

| Streamlit | Gradio | Static |

Figure 5-2. *Create a new space*

Provide a name for your space and choose Gradio as the SDK. Keep
public as default for now for visibility and finally click the Create
Space button.

You will see the following menu as shown in Figure 5-3.

▦ Spaces: ● smjain / **nlp** ▭   ♡ like 0   ● No application file

● App   ▪≡ Files and versions   ☁ Community   ⚙ Settings

Figure 5-3. *Menu on display on the huggingface web page*

For most of our applications in this chapter, we will use the Files and
versions and App tabs.

Click the Files and versions tab, and on the right side, we see Add file. Clicking it we can add the files needed for our application.

For our application we need only two files that we need to create:

1. app.py: This is the file that is the main code of the Gradio application.

2. requirements.txt: This file has the Python dependencies needed for the app.

Hugging Face Tasks

We will start with a question and answering task.

Question and Answering

The input to the model would be a paragraph and a question from within that paragraph. The output of the model inference would be the answer to the question.

The models we use are trained on the SQuAD dataset.

The Stanford Question Answering Dataset, also known as SQuAD, is a reading comprehension dataset that is comprised of questions that were posed by crowdworkers on a set of Wikipedia articles. The answer to every question is a segment of text, also known as a span, from the corresponding reading passage, or the question may be unanswerable.

More than 100,000 question and answer pairs are included in SQuAD 1.1, which covers more than 500 different articles.

First, the RoBERTa base model is used, fine-tuned using the SQuAD 2.0 dataset. It's been trained on question-answer pairs, including unanswerable questions, for the task of question and answering.

Some of the hyperparameters used by the model are

batch_size: 96

n_epochs: 2

max_seq_len: 386

max_query_length: 64

Start by creating a **new space** using the huggingface UI as explained in steps in the previous section.

Click the Files and versions tab on the UI. Create a file requirements.txt with the following content:

requirements.txt

gradio

transformers

torch

Create another file app.py and copy the content from Listing 5-1.

Listing 5-1. Code for app.py

```
from transformers import AutoModelForQuestionAnswering,
AutoTokenizer, pipeline
import gradio as grad
import ast
mdl_name = "deepset/roberta-base-squad2"
my_pipeline = pipeline('question-answering', model=mdl_name,
tokenizer=mdl_name)

def answer_question(question,context):
    text= "{"+"'question': '"+question+"','context':
    '"+context+"'}"

    di=ast.literal_eval(text)
    response = my_pipeline(di)
    return response
grad.Interface(answer_question, inputs=["text","text"],
outputs="text").launch()
```

Commit the changes by clicking the Commit changes button as shown in Figure 5-4.

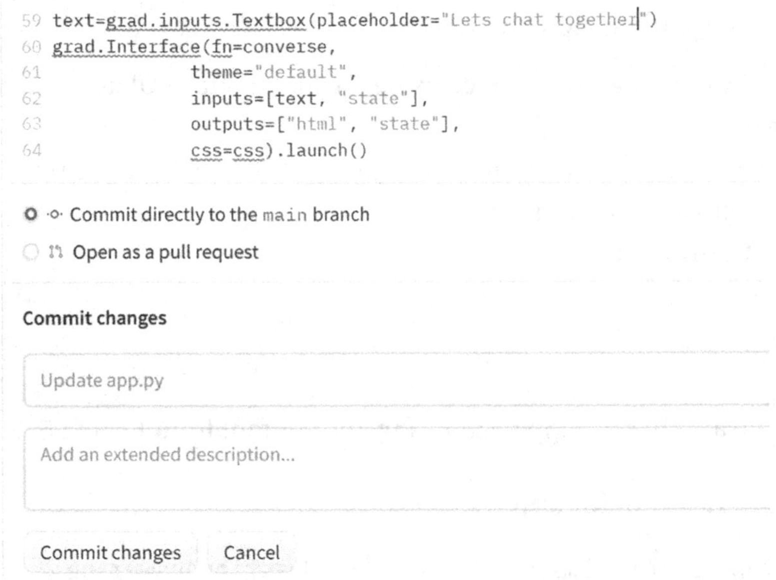

```
59 text=grad.inputs.Textbox(placeholder="Lets chat together")
60 grad.Interface(fn=converse,
61                theme="default",
62                inputs=[text, "state"],
63                outputs=["html", "state"],
64                css=css).launch()
```

○ ⦿ Commit directly to the main branch

○ ⇅ Open as a pull request

Commit changes

Update app.py

Add an extended description...

Commit changes   Cancel

Figure 5-4. *Commit the app.py file*

This would trigger the build and deployment process, and one can click the See logs button as in Figure 5-5 to see the activity.

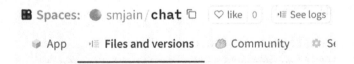

Figure 5-5. *Shows the various tabs including the "See logs" button*

The initial stage will be the building stage as shown in Figure 5-6.

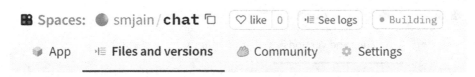

Figure 5-6. *Status of the deployment of the app*

Clicking See logs we can see the activity as shown in Figure 5-7.

```
--> FROM docker.io/library/python:3.8.9@sha256:49d05fff9cb3b185b15ffd92d8e6bd61c20aa91613
DONE 0.0s

--> RUN pip install          datasets          huggingface-hub
CACHED

--> RUN mkdir app
CACHED

--> RUN apt-get update && apt-get install -y          ffmpeg          libsm6          libxex
CACHED

--> RUN pip install pip==22.0.2
CACHED

--> WORKDIR /home/user
CACHED

--> COPY packages.txt /root/packages.txt
CACHED

--> WORKDIR /home/user/app
CACHED

--> RUN useradd -m -u 1000 user
CACHED

--> RUN pip install -r requirements.txt
CACHED
```

Figure 5-7. *Shows the build progress of the app. Here, it's loading the Docker images for creating a container*

One can see here the Docker image is being built, and then it will be deployed. If everything runs successfully, we will see a green stage on the UI with status Running as shown in Figure 5-8.

⊞ Spaces: ● smjain / **chat** 🗗   ♡ like   0   ∙≣ See logs   ● Running

🛢 App   ∙≣ **Files and versions**   🕸 Community   ⚙ Settings

⑂ main ▾   chat / app.py

● smjain   Update app.py   515c60f

</> raw   🕐 history   ☺ blame   ✐ edit   🗑 delete

Figure 5-8. *Status of the app is changed to Running now*

Once this is done, click the App tab, which is to the left of the Files and versions tab. This would present you the UI for keying in the inputs. Once inputs are provided, please click the Submit button as shown in Figure 5-9.

Figure 5-9. *Question and answering UI via Gradio. Provide the input of your choice for the paragraph in the input box labeled context, and the question from that paragraph should be put in the input box labeled question*

In Listing 5-2, we will try the same paragraph and question on a different model. The model we'll use is `distilbert-base-cased-distilled-squad`:

requirements.txt
gradio
transformers
torch

Listing 5-2. Code for app.py

```
from transformers import AutoModelForQuestionAnswering,
AutoTokenizer, pipeline
import gradio as grad
import ast

mdl_name = "distilbert-base-cased-distilled-squad"
my_pipeline = pipeline('question-answering', model=mdl_name,
tokenizer=mdl_name)

def answer_question(question,context):
    text= "{"+"'question': '"+question+"','context':
    '"+context+"'}"

    di=ast.literal_eval(text)
    response = my_pipeline(di)
    return response
grad.Interface(answer_question, inputs=["text","text"],
outputs="text").launch()
```

Commit the changes and wait for the status of deployment to get green. Post that click the App tab in the menu to launch the application.

Provide inputs to the UI and click the Submit button to see the results as shown in Figure 5-10.

question		output
Who administers Delhi		{'score': 0.5966600775718689, 'start': 315, 'end': 319, 'answer': 'NDMC'}

context

New Delhi is the capital of India and a part of the National Capital
Territory of Delhi (NCT). New Delhi is the seat of all three branches of the
government of India, hosting the Rashtrapati Bhavan, Parliament House,
and the Supreme Court of India. New Delhi is a municipality within the
NCT, administrated by the NDMC, which covers mostly Lutyens Delhi and
a few adjacent areas. The municipal area is part of a larger administrative
district, the New Delhi district.

Clear	Submit

Figure 5-10. Shows the Q&A Gradio-based UI for BERT-based Q&A

Translation

The next task we will tackle is language translation. The idea behind this is
to take input in one language and translate it to another language based on
pretrained models loaded via the huggingface library.

The first model we explore here is the Helsinki-NLP/opus-mt-en-de
model, which takes input in English and translates it to German.

Code

app.py

Listing 5-3. Code for app.py

```python
from transformers import pipeline
import gradio as grad
mdl_name = "Helsinki-NLP/opus-mt-en-de"
opus_translator = pipeline("translation", model=mdl_name)

def translate(text):

    response = opus_translator(text)
```

```
    return response
grad.Interface(translate, inputs=["text",], outputs="text").
launch()
```

requirements.txt

gradio

transformers

torch

transformers[sentencepiece]

Output

Commit the changes and wait for the status of deployment to get green. Post that click the App tab in the menu to launch the application.

Provide inputs to the UI and click the Submit button to see the results as shown in Figure 5-11.

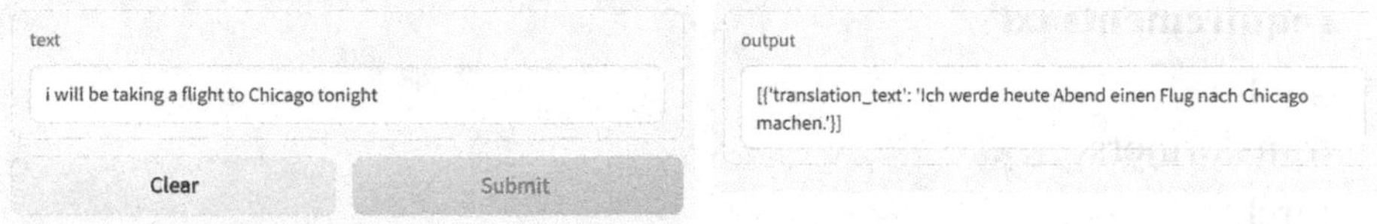

Figure 5-11. *Gradio UI for a translation task*

We will now see in Listing 5-4 if we can write the same code without using the pipeline abstraction. If we remember we have done the same earlier using the Auto classes like AutoTokenizer and AutoModel. Let's go ahead.

Code

app.py

Listing 5-4. Code for app.py

```
from transformers import AutoModelForSeq2SeqLM,AutoTokenizer
import gradio as grad
mdl_name = "Helsinki-NLP/opus-mt-en-de"
```

```
mdl = AutoModelForSeq2SeqLM.from_pretrained(mdl_name)
my_tkn = AutoTokenizer.from_pretrained(mdl_name)

#opus_translator = pipeline("translation", model=mdl_name)

def translate(text):
    inputs = my_tkn(text, return_tensors="pt")
    trans_output = mdl.generate(**inputs)
    response = my_tkn.decode(trans_output[0], skip_special_
    tokens=True)

    #response = opus_translator(text)
    return response
grad.Interface(translate, inputs=["text",], outputs="text").
launch()
```

requirements.txt
gradio
transformers
torch
transformers[sentencepiece]

Commit the changes and wait for the status of deployment to get green. Post that click the App tab in the menu to launch the application.

Provide inputs to the UI and click the Submit button to see the results as shown in Figure 5-12.

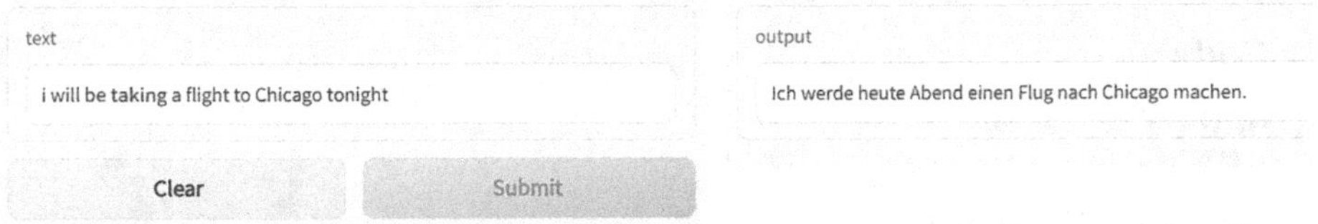

text	output
i will be taking a flight to Chicago tonight	Ich werde heute Abend einen Flug nach Chicago machen.

| Clear | Submit |

Figure 5-12. *Translation UI based on Gradio*

To give you a feeling of happiness, when we try the same translation via Google Translate, we get the following results as shown in Figure 5-13.

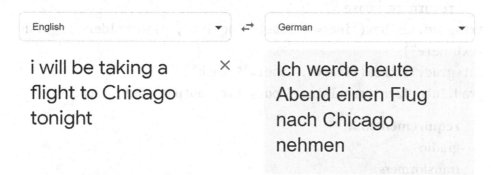

Figure 5-13. *Shows how Google Translate translates the same text we used for our translation app*

We can see how close we are to Google's results. This is the power of huggingface models.

To reinforce the concept, we will repeat the exercise with a different language translation. This time we take an example of English-to-French translation.

This time we take the Helsinki-NLP/opus-mt-en-fr model and try translating the same sentence we took in the preceding example, but this time to French.

First, we write the code using the pipeline abstraction.

Code

app.py

Listing 5-5. Code for app.py

```
from transformers import pipeline
import gradio as grad
mdl_name = "Helsinki-NLP/opus-mt-en-fr"

opus_translator = pipeline("translation", model=mdl_name)
```

```
def translate(text):
    response = opus_translator(text)
    return response
txt=grad.Textbox(lines=1, label="English", placeholder="English
Text here")
out=grad.Textbox(lines=1, label="French")
grad.Interface(translate, inputs=txt, outputs=out).launch()
```

requirements.txt

gradio

transformers

torch

transformers[sentencepiece]

Commit the changes and wait for the status of deployment to get green. Post that click the App tab in the menu to launch the application.

Provide inputs to the UI and click the Submit button to see the results as shown in Figure 5-14.

We get the following output.

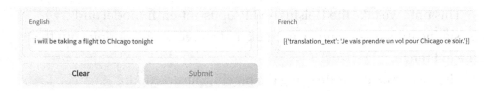

Figure 5-14. Translation UI using Gradio

Next, we try the same without the pipeline API in Listing 5-6.

Code

app.py

Listing 5-6. Code for app.py

```
from transformers import AutoModel,AutoTokenizer,AutoModelFor
Seq2SeqLM
import gradio as grad
mdl_name = "Helsinki-NLP/opus-mt-en-fr"
mdl = AutoModelForSeq2SeqLM.from_pretrained(mdl_name)
my_tkn = AutoTokenizer.from_pretrained(mdl_name)

#opus_translator = pipeline("translation", model=mdl_name)

def translate(text):
    inputs = my_tkn(text, return_tensors="pt")
    trans_output = mdl.generate(**inputs)
    response = my_tkn.decode(trans_output[0], skip_special_
    tokens=True)

    #response = opus_translator(text)
    return response
txt=grad.Textbox(lines=1, label="English", placeholder="English
Text here")
out=grad.Textbox(lines=1, label="French")
grad.Interface(translate, inputs=txt, outputs=out).launch()
```

requirements.txt

gradio

transformers

torch

transformers[sentencepiece]

Commit the changes and wait for the status of deployment to get green. Post that click the App tab in the menu to launch the application.

Provide inputs to the UI and click the Submit button to see the results as shown in Figure 5-15.

Figure 5-15. *Gradio UI for a translation task without using the pipeline API directly*

Again, compare the results with those of Google Translate, as shown in Figure 5-16.

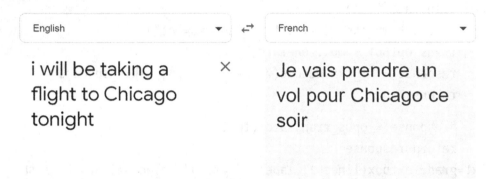

Figure 5-16. *Google Translate for the same text we used for our Gradio app*

As we can see, the results are exactly matching. Something to cheer about.

Summary

If we are presented with lengthy documents to read, our natural inclination is to either not read them at all or skim only the most important points. Therefore, it would be very helpful to have a summary of the information to save both time and mental processing power.

In the past, however, automatically summarizing text was an impossible task. To be more specific, making an abstractive summary is a very difficult task. Abstractive summarization is more difficult than extractive summarization, which pulls key sentences from a document and combines them to form a "summary." Because abstractive summarization involves paraphrasing words, it is also more time-consuming; however, it has the potential to produce a more polished and coherent summary.

We will be first looking at the google/pegasus-xsum model to generate the summary of some text.

Here is the code.

app.py

Listing 5-7. Code for app.py

```
from transformers import PegasusForConditionalGeneration,
PegasusTokenizer
import gradio as grad
mdl_name = "google/pegasus-xsum"
pegasus_tkn = PegasusTokenizer.from_pretrained(mdl_name)
mdl = PegasusForConditionalGeneration.from_pretrained(mdl_name)

def summarize(text):
    tokens = pegasus_tkn(text, truncation=True,
    padding="longest", return_tensors="pt")
    txt_summary = mdl.generate(**tokens)
    response = pegasus_tkn.batch_decode(txt_summary, skip_
    special_tokens=True)
    return response
txt=grad.Textbox(lines=10, label="English",
placeholder="English Text here")
out=grad.Textbox(lines=10, label="Summary")
grad.Interface(summarize, inputs=txt, outputs=out).launch()
```

requirements.txt

gradio

transformers

torch

transformers[sentencepiece]

Commit the changes and wait for the status of deployment to get green. Post that click the App tab in the menu to launch the application.

Provide inputs to the UI and click the Submit button to see the results as shown in Figure 5-17.

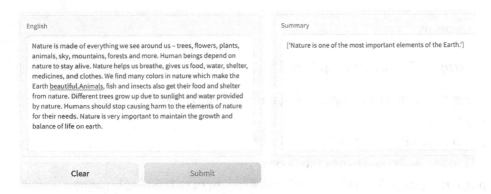

Figure 5-17. *Summarization app using Gradio. Paste a para in the box labeled English, and upon submitting, the box labeled Summary will show the summary of the para text*

Next, we take another text and also apply some tuning to the model with some parameters.

Listing 5-8. Code for app.py

```
from transformers import PegasusForConditionalGeneration,
PegasusTokenizer
import gradio as grad
mdl_name = "google/pegasus-xsum"
pegasus_tkn = PegasusTokenizer.from_pretrained(mdl_name)
```

```
mdl = PegasusForConditionalGeneration.from_pretrained(mdl_name)

def summarize(text):
    tokens = pegasus_tkn(text, truncation=True,
    padding="longest", return_tensors="pt")
    translated_txt = mdl.generate(**tokens,num_return_
    sequences=5,max_length=200,temperature=1.5,num_beams=10)
    response = pegasus_tkn.batch_decode(translated_txt, skip_
    special_tokens=True)
    return response
txt=grad.Textbox(lines=10, label="English",
placeholder="English Text here")
out=grad.Textbox(lines=10, label="Summary")
grad.Interface(summarize, inputs=txt, outputs=out).launch()
```

Commit the changes and wait for the status of deployment to get green. Post that click the App tab in the menu to launch the application.

Provide inputs to the UI and click the Submit button to see the results as shown in Figure 5-18.

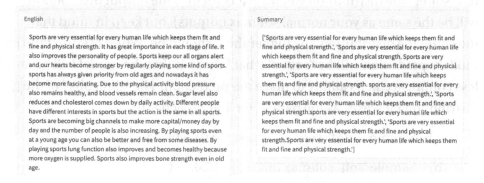

Figure 5-18. *Summary of the text via the Gradio app*

We can see we provided the following parameters in the code:

```
translated_txt = mdl.generate(**tokens,num_return_
sequences=5,max_length=200,temperature=1.5,num_beams=10)
```

Text generation is accomplished through the use of beam search, which is what num beams refers to. In contrast with greedy search, which only returns the next word that is most likely to be found, this method returns the n words that are most likely to be found.

Num_return_sequences returns the number of outputs returned. In the preceding example, we gave 5 as the number of sequences.

Altering the output distribution that is computed by your neural network is one justification for making use of the temperature function. In accordance with Equation 5-1 (temperature function), it is added to the logits vector:

$$qi = \exp(zi/T)/ \sum\nolimits_j \exp(zj/T) \qquad \text{Equation 5-1}$$

where T is the temperature parameter.

You must understand that this will result in a shift in the overall probabilities. You are free to pick any value for T you like (the higher the T, the "softer" the distribution will be; if it is set to 1, the output distribution will be the same as your normal softmax outputs), but keep in mind that the softer the distribution, the higher the T should be. When I say that the model's prediction will be "softer," what I mean is that the model will have less confidence in its ability to make the prediction. The "difficulty" of the distribution increases as the parameter T approaches 0.

a)   Sample "hard" softmax probs : [0.01,0.04,0.95]

b)   Sample 'soft' softmax probs : [0.15,0.25,0.6]

In the preceding "a" is a harder distribution. Your model exhibits a high level of assurance regarding its predictions. On the other hand, you probably don't want your model to behave in that way in most situations. For instance, if you are generating text with an RNN, you are essentially taking a sample from your output distribution and using that sampled word as your output token (and next input). If your model has a high level of self-assurance, it may generate text that is very repetitive and not very interesting. You want it to produce text with a greater variety of topics, but it won't be able to do that because, while the sampling procedure is taking place, the majority of the probability mass will be concentrated in a few tokens, and as a result, your model will keep selecting the same small group of words over and over again. You could plug in the temperature variable to generate more diverse text and give other words a chance to be sampled as well. This would accomplish the same thing.

The exponential function is to blame for the fact that higher temperatures result in flatter distributions. This is because of how the function works. The temperature parameter places a greater amount of negative weight on larger logits than it does on smaller logits. An "increasing function" is what the exponential function is called. Therefore, if a term is already quite significant, penalizing it by a small amount would make it much smaller (in terms of percentage) than if that term was relatively minor.

For a more keen user, here is a brief about the Pegasus model.

Pegasus

During the pretraining phase of the Pegasus system, several complete sentences are deleted from the source document. The model is then tasked with retrieving these sentences. The missing sentences from the document serve as the input for such pretraining, while the document itself serves as the output. The input document is what is known as the "input document." This is a self-supervised model without any need of annotations in the training dataset.

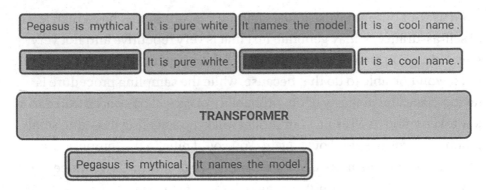

Figure 5-19. *Google Pegasus model for summarizing text*
Imagsource: `https://1.bp.blogspot.com/-TS0r4o51jGI/`
`Xt5Olkj6blI/AAAAAAAAGDs/`

Zero-Shot Learning

Zero-shot learning, as the name implies, is to use a pretrained model, trained on a certain set of data, on a different set of data, which it has not seen during training. This would mean, as an example, to take some model from huggingface that is trained on a certain dataset and use it for inference on examples it has never seen before.

Zero-Shot Text Classification

Text classification is a task in natural language processing that requires the model to make predictions about the classes that the text documents belong to, as is common knowledge. In the traditional method, we are required to use a significant amount of data that has been labeled in order to train the model. Additionally, they are unable to make predictions using data that they have not seen. The use of zero-shot learning in conjunction with text classification has reached an unprecedented level of sophistication in natural language processing.

The primary objective of any model associated with the zero-shot text classification technique is to classify the text documents without employing any labeled data or without having seen any labeled text. This can be accomplished by classifying the documents without having seen any labeled text. The transformers are where the zero-shot classification implementations are most frequently found by us. There are more than 60 transformer models that function based on the zero-shot classification that are found in the huggingface library.

When we discuss zero-shot text classification, there is one additional thing that springs to mind. In the same vein as zero-shot classification is few-shot classification, which is very similar to zero-shot classification. However, in contrast with zero-shot classification, few-shot classification makes use of very few labeled samples during the training process. The implementation of the few-shot classification methods can be found in OpenAI, where the GPT3 classifier is a well-known example of a few-shot classifier.

Why We Need Zero-Shot

1. There is either no data at all or only a very limited amount of data available for training (detection of the user's intentions without receiving any data from the user).

2. There are an extremely high number of classes and labels (thousands upon thousands).

3. A classifier that works "out of the box," with reduced costs for both infrastructure and development.

We will use the pipeline API first to see if we can just create a simple classifier with zero-shot learning.

Code

app.py

Listing 5-9. Code for app.py

```
from transformers import pipeline
import gradio as grad
zero_shot_classifier = pipeline("zero-shot-classification")

def classify(text,labels):
    classifer_labels = labels.split(",")
    #["software", "politics", "love", "movies", "emergency",
    "advertisment","sports"]
    response = zero_shot_classifier(text,classifer_labels)
    return response
txt=grad.Textbox(lines=1, label="English", placeholder="text to
be classified")
labels=grad.Textbox(lines=1, label="Labels", placeholder="comma
separated labels")
out=grad.Textbox(lines=1, label="Classification")
grad.Interface(classify, inputs=[txt,labels], outputs=out).
launch()
```

requirements.txt
gradio
transformers
torch
transformers[sentencepiece]

Commit the changes and wait for the status of deployment to get green. Post that click the App tab in the menu to launch the application.

Provide inputs to the UI and click the Submit button to see the results as shown in Figure 5-20.

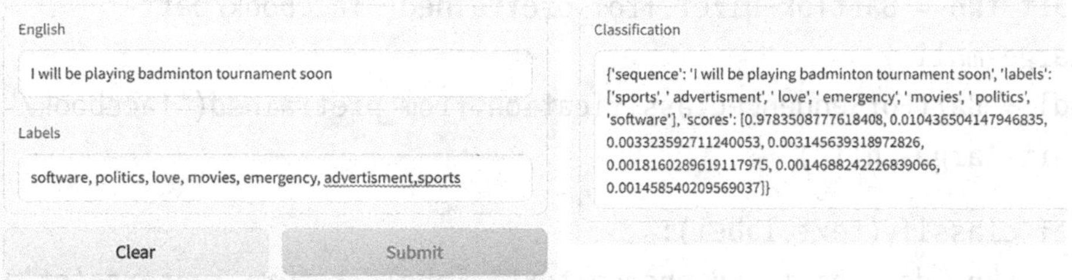

Figure 5-20. *Zero-shot classification*

We can see this text gets classified under the sports category correctly. We try another example as shown in Figure 5-21.

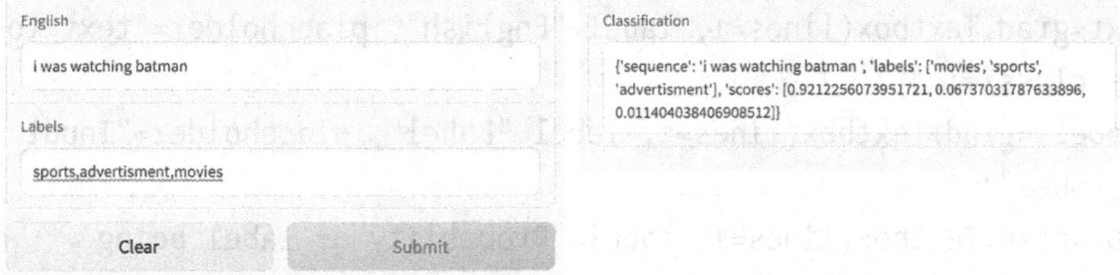

Figure 5-21. *Another example of zero-shot classification. The box labeled Classification represents the scores/probabilities of individual classes*

Another way to attain the same without the pipeline API is shown in the following.

Code

app.py

Listing 5-10. Code for app.py

```
from transformers import BartForSequenceClassification,
BartTokenizer
import gradio as grad
```

```
bart_tkn = BartTokenizer.from_pretrained('facebook/bart-
large-mnli')
mdl = BartForSequenceClassification.from_pretrained('facebook/
bart-large-mnli')

def classify(text,label):
    tkn_ids = bart_tkn.encode(text, label, return_tensors='pt')
    tkn_lgts = mdl(tkn_ids)[0]
    entail_contra_tkn_lgts = tkn_lgts[:,[0,2]]
    probab = entail_contra_tkn_lgts.softmax(dim=1)
    response =  probab[:,1].item() * 100
    return response
txt=grad.Textbox(lines=1, label="English", placeholder="text to
be classified")
labels=grad.Textbox(lines=1, label="Label", placeholder="Input
a Label")
out=grad.Textbox(lines=1, label="Probablity of label being
true is")
grad.Interface(classify, inputs=[txt,labels], outputs=out).
launch()
```

requirements.txt
gradio
transformers
torch

Commit the changes and wait for the status of deployment to get green. Post that click the App tab in the menu to launch the application.

Provide inputs to the UI and click the Submit button to see the results as shown in Figure 5-22.

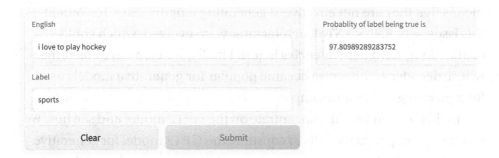

Figure 5-22. *Gradio app for zero-shot classification*

This again classifies the text under the sports category correctly.

Text Generation Task/Models

The development of text generation models started many decades ago, which is a long time before the recent surge in interest in deep learning. Given a piece of text, models of this kind should be able to make accurate predictions about a particular word or string of words. Given a text as input, the model navigates through a search space to generate probabilities of what could be the next probable word from a joint distribution of words.

The earliest text generation models were based on Markov chains. Markov chains are like a state machine wherein using only the previous state, the next state is predicted. This is similar also to what we studied in bigrams.

Post the Markov chains, recurrent neural networks (RNNs), which were capable of retaining a greater context of the text, were introduced. They are based on neural network architectures that are recurrent in nature. RNNs are able to retain a greater context of the text that was introduced. Nevertheless, the amount of information that these kinds of networks are able to remember is constrained, and it is also difficult to train them, which means that they are not effective at generating lengthy texts. To counter this issue with RNNs, LSTM architectures were evolved, which could capture long-term dependencies in text. Finally, we came to transformers, whose decoder architecture became popular for generative models used for generating text as an example.

In this section we will concentrate on the GPT2 model and see how we can use the huggingface APIs to consume the GPT2 model for generative tasks. This will allow us to generate text with the pretrained models and also fine-tune them if needed with a custom text dataset.

Code

app.py

Listing 5-11. Code for app.py

```python
from transformers import GPT2LMHeadModel,GPT2Tokenizer
import gradio as grad

mdl = GPT2LMHeadModel.from_pretrained('gpt2')
gpt2_tkn=GPT2Tokenizer.from_pretrained('gpt2')

def generate(starting_text):
    tkn_ids = gpt2_tkn.encode(starting_text, return_
    tensors = 'pt')
    gpt2_tensors = mdl.generate(tkn_ids)
    response = gpt2_tensors
    return response
txt=grad.Textbox(lines=1, label="English", placeholder="English
Text here")
out=grad.Textbox(lines=1, label="Generated Tensors")
grad.Interface(generate, inputs=txt, outputs=out).launch()
```

requirements.txt
gradio

transformers

torch

Commit the changes and wait for the status of deployment to get green. Post that click the App tab in the menu to launch the application.

Provide inputs to the UI and click the Submit button to see the results as shown in Figure 5-23.

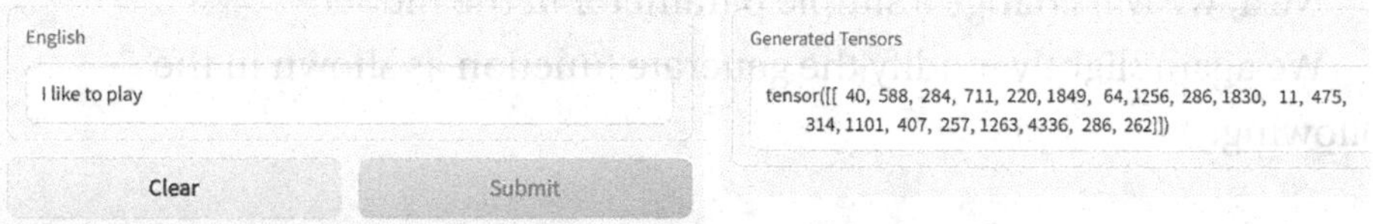

English	Generated Tensors
I like to play	tensor([[40, 588, 284, 711, 220, 1849, 64, 1256, 286, 1830, 11, 475, 314, 1101, 407, 257, 1263, 4336, 286, 262]])

Clear	Submit

Figure 5-23. *Tensors for the English text*

Next, we decode these tensors in the generate function.

Listing 5-12. Generate function

```
def generate(starting_text):
    tkn_ids = gpt2_tkn.encode(starting_text, return_
    tensors = 'pt')
    gpt2_tensors = mdl.generate(tkn_ids)
    response=""
    #response = gpt2_tensors
    for i, x in enumerate(gpt2_tensors):
        response=response+f"{i}: {gpt2_tkn.decode(x, skip_
        special_tokens=True)}"
    return response
```

I will give another text as input and check.

Commit the changes and wait for the status of deployment to get green. Post that click the App tab in the menu to launch the application.

Provide inputs to the UI and click the Submit button to see the results as shown in Figure 5-24.

Figure 5-24. *Code generation app via Gradio*

Next, we will change a simple parameter in the model.

We again slightly modify the generate function as shown in the following.

Listing 5-13. Modified code for the generate function of app.py

```
def generate(starting_text):
    tkn_ids = gpt2_tkn.encode(starting_text, return_
    tensors = 'pt')
    gpt2_tensors = mdl.generate(tkn_ids,max_length=100)
    response=""
    #response = gpt2_tensors
    for i, x in enumerate(gpt2_tensors):
        response=response+f"{i}: {gpt2_tkn.decode(x, skip_
        special_tokens=True)}"
    return response
```

Commit the changes and wait for the status of deployment to get green. Post that click the App tab in the menu to launch the application.

Provide inputs to the UI and click the Submit button to see the results as shown in Figure 5-25.

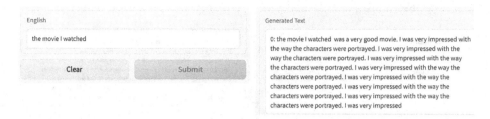

Figure 5-25. *Text generation using Gradio*

We can see there is a lot of repetition happening in the generated output. To mitigate this we add another parameter to the model.

Listing 5-14. Modified code for the generate function of app.py to avoid repetition

```
def generate(starting_text):
    tkn_ids = gpt2_tkn.encode(starting_text, return_
    tensors = 'pt')
    gpt2_tensors = mdl.generate(tkn_ids,max_length=100,no_
    repeat_ngram_size=True)
    response=""
    #response = gpt2_tensors
    for i, x in enumerate(gpt2_tensors):
        response=response+f"{i}: {gpt2_tkn.decode(x, skip_
        special_tokens=True)}"
    return response
```

Commit the changes and wait for the status of deployment to get green. Post that click the App tab in the menu to launch the application.

Provide inputs to the UI and click the Submit button to see the results as shown in Figure 5-26.

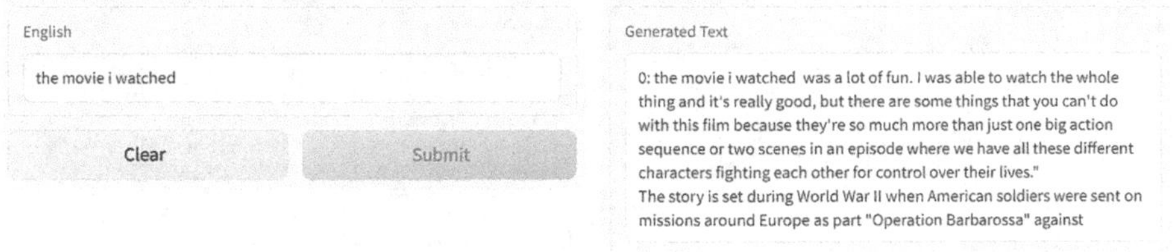

English	Generated Text
the movie i watched	0: the movie i watched was a lot of fun. I was able to watch the whole thing and it's really good, but there are some things that you can't do with this film because they're so much more than just one big action sequence or two scenes in an episode where we have all these different characters fighting each other for control over their lives." The story is set during World War II when American soldiers were sent on missions around Europe as part "Operation Barbarossa" against

Clear   *Submit*

Figure 5-26. *Text generation where the generator now avoids repetition in text*

So far the search being done by the model to find the next word is based on greedy search.

It is the most straightforward approach, which entails selecting the word from all of the alternatives that has the highest probability of being correct. It is the one that is used whenever there is no specific parameter specified. This process is deterministic in nature, which means that resultant text is the same as before if we use the same parameters.

Next, we specify a parameter num_beams to perform a beam search.

It returns the sequences that have the highest probability, and then, when it comes time to choose, it picks the one that has the highest probability. The value of num_beams is represented by the parameter X. We again modify the generate function to adjust this parameter.

Listing 5-15. Modified code for the generate function of app.py, num_beams specified

```
def generate(starting_text):
    tkn_ids = gpt2_tkn.encode(starting_text, return_
    tensors = 'pt')
    gpt2_tensors = mdl.generate(tkn_ids,max_length=100,
    no_repeat_ngram_size=True,num_beams=3)
    response=""
    #response = gpt2_tensors
    for i, x in enumerate(gpt2_tensors):
```

```
    response=response+f"{i}: {gpt2_tkn.decode(x, skip_
    special_tokens=True)}"
  return response
```

Commit the changes and wait for the status of deployment to get green. Post that click the App tab in the menu to launch the application.

Provide inputs to the UI and click the Submit button to see the results as shown in Figure 5-27.

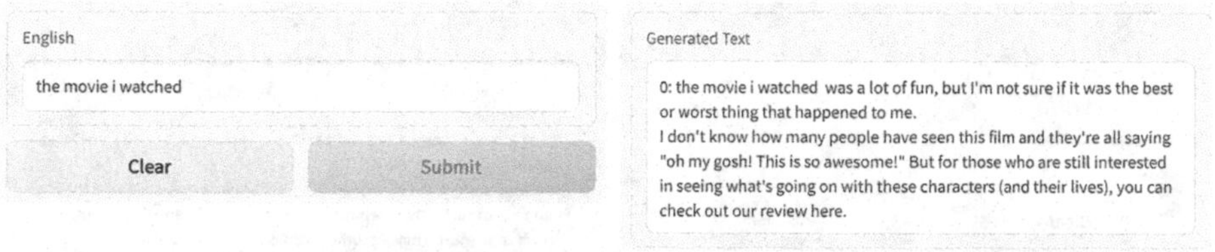

English

the movie i watched

Clear   Submit

Generated Text

0: the movie i watched was a lot of fun, but I'm not sure if it was the best or worst thing that happened to me.
I don't know how many people have seen this film and they're all saying "oh my gosh! This is so awesome!" But for those who are still interested in seeing what's going on with these characters (and their lives), you can check out our review here.

Figure 5-27. *UI showing text generation after using beam search in the generate function*

The next approach we take is sampling.

Sampling is a parameter by which the next word is selected at random from the probability distribution.

In this case we set the parameter do_sample=true inside the generate function.

Listing 5-16. Modified code for the generate function of app.py with sampling

```
def generate(starting_text):
    tkn_ids = gpt2_tkn.encode(starting_text, return_
    tensors = 'pt')
    gpt2_tensors = mdl.generate(tkn_ids,max_length=100,no_
    repeat_ngram_size=True,num_beams=3,do_sample=True)
    response=""
    #response = gpt2_tensors
```

```
for i, x in enumerate(gpt2_tensors):
    response=response+f"{i}: {gpt2_tkn.decode(x, skip_
    special_tokens=True)}"
return response
```

Commit the changes and wait for the status of deployment to get green. Post that click the App tab in the menu to launch the application.

Provide inputs to the UI and click the Submit button to see the results as shown in Figure 5-28.

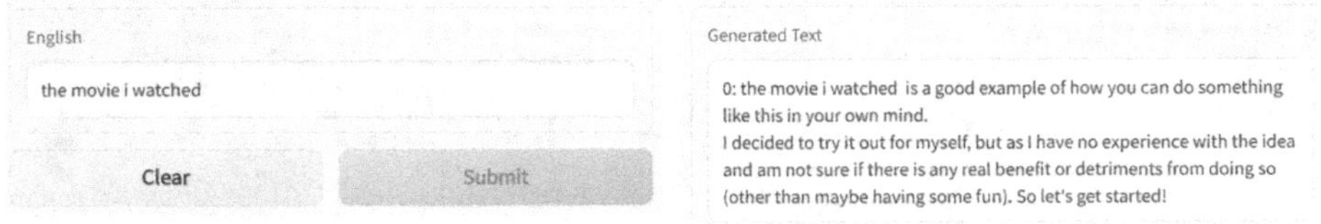

English

the movie i watched

Clear   Submit

Generated Text

0: the movie i watched is a good example of how you can do something like this in your own mind.
I decided to try it out for myself, but as I have no experience with the idea and am not sure if there is any real benefit or detriments from doing so (other than maybe having some fun). So let's get started!

Figure 5-28. Gradio UI with the generate function using sampling behind the scenes

It is possible to alter the "temperature" of the distribution in order to raise the likelihood of successfully removing a word from among the most likely candidates.

The level of greed that the generative model exhibits is proportional to the temperature.

If the temperature is low, the probabilities of sample classes other than the one with the highest log probability will be low. As a result, the model will probably output the text that is most correct, but it will be rather monotonous and contain only a small amount of variation.

If the temperature is high, the model has a greater chance of outputting different words than those with the highest probability. The generated text will feature a greater variety of topics, but there is also an increased likelihood that it will generate nonsense text and contain grammatical errors.

We again modify the generate function.

Listing 5-17. Modified code for the generate function of app.py with temperature setting of 1.5

```
def generate(starting_text):
    tkn_ids = gpt2_tkn.encode(starting_text, return_
    tensors = 'pt')
    gpt2_tensors = mdl.generate(tkn_ids,max_
    length=100,no_repeat_ngram_size=True,num_beams=3,do_
    sample=True,temperatue=1.5)
    response=""
    #response = gpt2_tensors
    for i, x in enumerate(gpt2_tensors):
        response=response+f"{i}: {gpt2_tkn.decode(x, skip_
        special_tokens=True)}"
    return response
```

Commit the changes and wait for the status of deployment to get green. Post that click the App tab in the menu to launch the application.

Provide inputs to the UI and click the Submit button to see the results as shown in Figure 5-29.

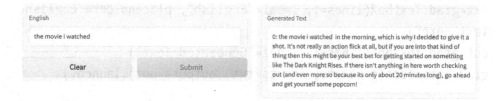

Figure 5-29. *Text generation shown in Gradio UI with the generate function using a temperature setting of 1.5*

When we run the same code with less temperature, the output becomes less variational.

Listing 5-18. Modified code for the generate function of app.py with temperature setting of 0.1

```
from transformers import GPT2LMHeadModel,GPT2Tokenizer
import gradio as grad

mdl = GPT2LMHeadModel.from_pretrained('gpt2')
gpt2_tkn=GPT2Tokenizer.from_pretrained('gpt2')

def generate(starting_text):
    tkn_ids = gpt2_tkn.encode(starting_text, return_
    tensors = 'pt')
    gpt2_tensors = mdl.generate(tkn_ids,max_
    length=100,no_repeat_ngram_size=True,num_beams=3,do_
    sample=True,temperatue=0.1)
    response=""
    #response = gpt2_tensors
    for i, x in enumerate(gpt2_tensors):
        response=response+f"{i}: {gpt2_tkn.decode(x, skip_
        special_tokens=True)}"
    return response
txt=grad.Textbox(lines=1, label="English", placeholder="English
Text here")
out=grad.Textbox(lines=1, label="Generated Text")
grad.Interface(generate, inputs=txt, outputs=out).launch()
```

Commit the changes and wait for the status of deployment to get green. Post that click the App tab in the menu to launch the application.

Provide inputs to the UI and click the Submit button to see the results as shown in Figure 5-30.

English	Generated Text
the movie i watched	0: the movie i watched was a good example of this. I have to say that it was really interesting, even if the film itself wasn't quite as great (for me at least).
Clear **Submit**	If you're wondering what's going on with Raging Bull, here is my review: http://i7-cdn2w8s1g6yv3p5jk4xmqd9c0n_uNzR/viewer?attachment=f

Figure 5-30. *Text generation shown in Gradio UI with the generate function using a temperature setting of 0.1*

To let the concepts of text generation sink in a bit more, we will take another model into consideration called the "distilgpt2."

Code

app.py

Listing 5-19. app.py code using the GPT2 model for text generation

```
from transformers import pipeline, set_seed
import gradio as grad
gpt2_pipe = pipeline('text-generation', model='distilgpt2')
set_seed(42)

def generate(starting_text):
    response= gpt2_pipe(starting_text, max_length=20, num_
    return_sequences=5)
    return response
txt=grad.Textbox(lines=1, label="English", placeholder="English
Text here")
out=grad.Textbox(lines=1, label="Generated Text")
grad.Interface(generate, inputs=txt, outputs=out).launch()
```

requirements.txt
gradio
transformers
torch
transformers[sentencepiece]

Commit the changes and wait for the status of deployment to get green. Post that click the App tab in the menu to launch the application.

Provide inputs to the UI and click the Submit button to see the results as shown in Figure 5-31.

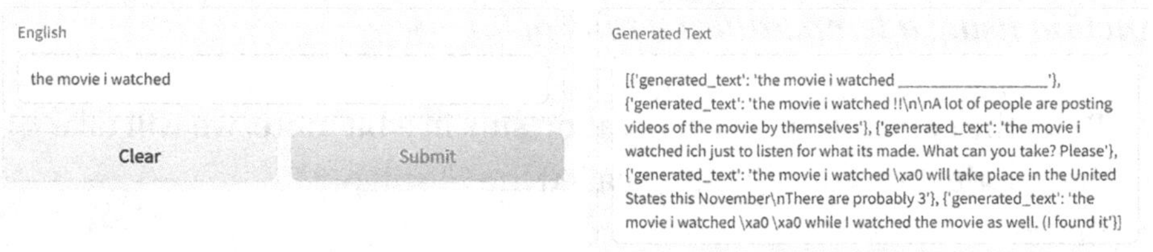

Figure 5-31. *Text generation using the GPT2 model behind the scenes*

Text-to-Text Generation

In this section, we will cover text-to-text generation using the T5 model.

A transformer-based architecture that takes a text-to-text approach is referred to as T5, which stands for Text-to-Text Transfer Transformer.

In the text-to-text approach, we take a task like Q&A, classification, summarization, code generation, etc. and turn it into a problem, which provides the model with some form of input and then teaches it to generate some form of target text. This makes it possible to apply the same model, loss function, hyperparameters, and other settings to all of our varied sets of responsibilities.

The T5 model, which was developed by Google Research and made public, contributes the following to previously conducted research:

1.   It produces a tidier version of the enormous Common Crawl dataset, which is referred to as the Colossal Cleaned Common Crawl (C4). This dataset is approximately 100,000 times more extensive than Wikipedia.

2. It prepares the body for T5 on the Common Crawl.

3. It proposes rethinking each and every NLP task as a formulation of an input text to an output text.

4. It demonstrates that state-of-the-art results can be achieved through fine-tuning on a variety of tasks, such as summarization, Q&A, and reading comprehension, by making use of the pretrained T5 and the text-to-text formulation.

5. Additionally, the T5 team conducted an in-depth study in order to learn the most effective methods for pretraining and fine-tuning. In their paper, they detail which parameters are most important to achieving desirable results.

Code

app.py

Listing 5-20. app.py code

```
from transformers import AutoModelWithLMHead, AutoTokenizer
import gradio as grad

text2text_tkn = AutoTokenizer.from_pretrained("mrm8488/t5-base-
finetuned-question-generation-ap")
mdl = AutoModelWithLMHead.from_pretrained("mrm8488/t5-base-
finetuned-question-generation-ap")

def text2text(context,answer):
    input_text = "answer: %s   context: %s </s>" % (answer,
    context)
    features = text2text_tkn ([input_text], return_
    tensors='pt')
```

```
output = mdl.generate(input_ids=features['input_ids'],
            attention_mask=features['attention_mask'],
            max_length=64)

response=text2text_tkn.decode(output[0])
return response
```

```
context=grad.Textbox(lines=10, label="English",
placeholder="Context")
ans=grad.Textbox(lines=1, label="Answer")
out=grad.Textbox(lines=1, label="Genereated Question")
grad.Interface(text2text, inputs=[context,ans], outputs=out).
launch()
```

requirements.txt
gradio
transformers
torch
transformers[sentencepiece]

Commit the changes and wait for the status of deployment to get green. Post that click the App tab in the menu to launch the application.

Provide inputs to the UI and click the Submit button to see the results as shown in Figure 5-32.

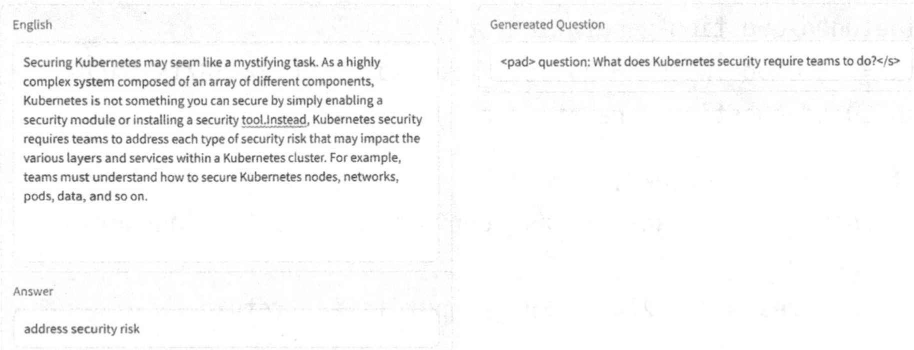

English

Securing Kubernetes may seem like a mystifying task. As a highly complex system composed of an array of different components, Kubernetes is not something you can secure by simply enabling a security module or installing a security tool.Instead, Kubernetes security requires teams to address each type of security risk that may impact the various layers and services within a Kubernetes cluster. For example, teams must understand how to secure Kubernetes nodes, networks, pods, data, and so on.

Answer

address security risk

Genereated Question

<pad> question: What does Kubernetes security require teams to do?</s>

Figure 5-32. *Generating questions from the para*

In same example, let's change the answer a bit.

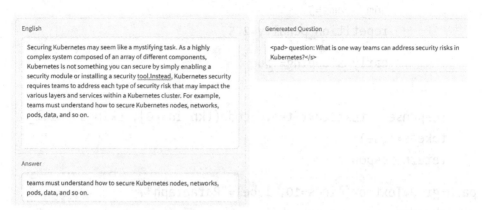

English

Securing Kubernetes may seem like a mystifying task. As a highly complex system composed of an array of different components, Kubernetes is not something you can secure by simply enabling a security module or installing a security tool.Instead, Kubernetes security requires teams to address each type of security risk that may impact the various layers and services within a Kubernetes cluster. For example, teams must understand how to secure Kubernetes nodes, networks, pods, data, and so on.

Genereated Question

<pad> question: What is one way teams can address security risks in Kubernetes?</s>

Answer

teams must understand how to secure Kubernetes nodes, networks, pods, data, and so on.

Figure 5-33. *Another example of question generation from text*

Now we look into another use case of T5, which is to summarize a paragraph of text.

Here is the code.

app.py

Listing 5-21. app.py code

```
from transformers import AutoTokenizer, AutoModelWithLMHead
import gradio as grad
text2text_tkn = AutoTokenizer.from_pretrained("deep-learning-
analytics/wikihow-t5-small")
mdl = AutoModelWithLMHead.from_pretrained("deep-learning-
analytics/wikihow-t5-small")

def text2text_summary(para):
    initial_txt = para.strip().replace("\n","")
    tkn_text = text2text_tkn.encode(initial_txt, return_
    tensors="pt")

    tkn_ids = mdl.generate(
            tkn_text,
```

```
            max_length=250,
            num_beams=5,
            repetition_penalty=2.5,

            early_stopping=True
        )

    response = text2text_tkn.decode(tkn_ids[0], skip_special_
    tokens=True)
    return response

para=grad.Textbox(lines=10, label="Paragraph",
placeholder="Copy paragraph")
out=grad.Textbox(lines=1, label="Summary")
grad.Interface(text2text_summary, inputs=para, outputs=out).
launch()
```

requirements.txt
gradio
transformers
torch
transformers[sentencepiece]

Commit the changes and wait for the status of deployment to get green. Post that click the App tab in the menu to launch the application.

Provide inputs to the UI and click the Submit button to see the results as shown in Figure 5-34.

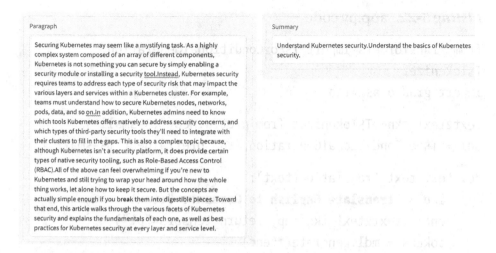

Paragraph	Summary
Securing Kubernetes may seem like a mystifying task. As a highly complex system composed of an array of different components, Kubernetes is not something you can secure by simply enabling a security module or installing a security tool.Instead, Kubernetes security requires teams to address each type of security risk that may impact the various layers and services within a Kubernetes cluster. For example, teams must understand how to secure Kubernetes nodes, networks, pods, data, and so on.In addition, Kubernetes admins need to know which tools Kubernetes offers natively to address security concerns, and which types of third-party security tools they'll need to integrate with their clusters to fill in the gaps. This is also a complex topic because, although Kubernetes isn't a security platform, it does provide certain types of native security tooling, such as Role-Based Access Control (RBAC).All of the above can feel overwhelming if you're new to Kubernetes and still trying to wrap your head around how the whole thing works, let alone how to keep it secure. But the concepts are actually simple enough if you break them into digestible pieces. Toward that end, this article walks through the various facets of Kubernetes security and explains the fundamentals of each one, as well as best practices for Kubernetes security at every layer and service level.	Understand Kubernetes security.Understand the basics of Kubernetes security.

Figure 5-34. *Summarizing text using the T5 model*

We now will take a few more tasks that we can perform using the T5 model. Some of them are listed in the following:

1. Translation

2. Sentiment classification

3. Paraphrasing

4. Classification of whether deduction of a statement from a sentence is right or not

5. And a few more

We will cover a few of the above-mentioned tasks in the following.

English-to-German Using T5

As highlighted in the following code segment, we need to prefix the text with `translate English to German` in order to generate the corresponding German translation.

Code

app.py

Listing 5-22. app.py code

```
from transformers import T5ForConditionalGeneration,
T5Tokenizer
import gradio as grad

text2text_tkn= T5Tokenizer.from_pretrained("t5-small")
mdl = T5ForConditionalGeneration.from_pretrained("t5-small")

def text2text_translation(text):
    inp = "translate English to German:: "+text
    enc = text2text_tkn(inp, return_tensors="pt")
    tokens = mdl.generate(**enc)
    response=text2text_tkn.batch_decode(tokens)
    return response

para=grad.Textbox(lines=1, label="English Text",
placeholder="Text in English")
out=grad.Textbox(lines=1, label="German Translation")
grad.Interface(text2text_translation, inputs=para,
outputs=out).launch()
```

requirements.txt
gradio
transformers
torch
transformers[sentencepiece]

Commit the changes and wait for the status of deployment to get green. Post that click the App tab in the menu to launch the application.

Provide inputs to the UI and click the Submit button to see the results as shown in Figure 5-35.

English Text

I like to play badminton

German Translation

['<pad> : Ich mag Badminton spielen</s>']

Clear   Submit

Figure 5-35. *Translation using the T5 model*

If we put this translated text in Google Translate, we get what is shown in Figure 5-36.

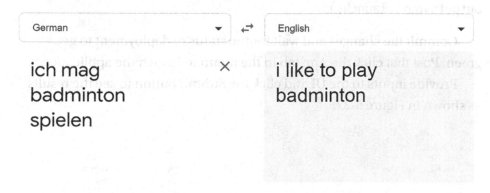

| German ▼ | ⇄ | English ▼ |

ich mag badminton spielen ✕

i like to play badminton

Figure 5-36. *Translation using Google Translate*

This is exactly the input we gave.

Now we change the prefix to translate English to French in app.py.

Listing 5-23. app.py code

```
from transformers import T5ForConditionalGeneration,
T5Tokenizer
import gradio as grad

text2text_tkn= T5Tokenizer.from_pretrained("t5-small")
mdl = T5ForConditionalGeneration.from_pretrained("t5-small")

def text2text_translation(text):
    inp = "translate English to French:: "+text
```

```
    enc = text2text_tkn(inp, return_tensors="pt")
    tokens = mdl.generate(**enc)
    response=text2text_tkn.batch_decode(tokens)
    return response

para=grad.Textbox(lines=1, label="English Text",
placeholder="Text in English")
out=grad.Textbox(lines=1, label="French Translation")
grad.Interface(text2text_translation, inputs=para,
outputs=out).launch()
```

Commit the changes and wait for the status of deployment to get green. Post that click the App tab in the menu to launch the application.

Provide inputs to the UI and click the Submit button to see the results as shown in Figure 5-37.

English Text	French Translation
i like to play badminton	["<pad> : j'aime jouer badminton</s>"]
Clear	Submit

Figure 5-37. *Another example of T5 model–based translation*

We perform a check on Google for the same.

Figure 5-38. *Translation using Google Translate*

We can see the result is exactly as that in Google Translate.

Sentiment Analysis Task

Next, we try a sentiment classification using the T5 model.

We use the sst2 sentence prefix for doing the sentiment analysis.

Code

app.py

Listing 5-24. app.py code

```
from transformers import T5ForConditionalGeneration,
T5Tokenizer
import gradio as grad

text2text_tkn= T5Tokenizer.from_pretrained("t5-small")
mdl = T5ForConditionalGeneration.from_pretrained("t5-small")

def text2text_sentiment(text):
    inp = "sst2 sentence: "+text
    enc = text2text_tkn(inp, return_tensors="pt")
    tokens = mdl.generate(**enc)
```

115

```
    response=text2text_tkn.batch_decode(tokens)
    return response

para=grad.Textbox(lines=1, label="English Text",
placeholder="Text in English")
out=grad.Textbox(lines=1, label="Sentiment")
grad.Interface(text2text_sentiment, inputs=para, outputs=out).
launch()
```

requirements.txt
gradio
transformers
torch
transformers[sentencepiece]

Commit the changes and wait for the status of deployment to get green. Post that click the App tab in the menu to launch the application.

Provide inputs to the UI and click the Submit button to see the results as shown in Figure 5-39.

Figure 5-39.   *Sentiment analysis task using T5 – positive*

Let's run the code again on a different text .

We get the following output.

English Text	Sentiment
the bear market is spoiling my party	['<pad> negative</s>']

Clear	Submit

Figure 5-40. *Sentiment analysis task using T5 – negative*

Again you can observe how easy life becomes by just using these pretrained models on a variety of tasks.

Next we use the T5 model to check the grammatical acceptance of a text by using the cola sentence prefix as shown in the following.

Code

app.py

Listing 5-25. app.py code

```
from transformers import T5ForConditionalGeneration,
T5Tokenizer
import gradio as grad

text2text_tkn= T5Tokenizer.from_pretrained("t5-small")
mdl = T5ForConditionalGeneration.from_pretrained("t5-small")
def text2text_acceptable_sentence(text):
    inp = "cola sentence: "+text
    enc = text2text_tkn(inp, return_tensors="pt")
    tokens = mdl.generate(**enc)
    response=text2text_tkn.batch_decode(tokens)
    return response

para=grad.Textbox(lines=1, label="English Text",
placeholder="Text in English")
out=grad.Textbox(lines=1, label="Whether the sentence is
acceptable or not")
```

```
grad.Interface(text2text_acceptable_sentence, inputs=para,
outputs=out).launch()
```

requirements.txt

gradio

transformers

torch

transformers[sentencepiece]

Commit the changes and wait for the status of deployment to get green. Post that click the App tab in the menu to launch the application.

Provide inputs to the UI and click the Submit button to see the results as shown in Figure 5-41.

Figure 5-41. *Sentence acceptability*

Sentence Paraphrasing Task

Now we check whether two sentences are paraphrases of each other using the mrpc sentence1 sentence2 prefix.

Code

app.py

Listing 5-26. app.py code

```
from transformers import T5ForConditionalGeneration,
T5Tokenizer
import gradio as grad

text2text_tkn= T5Tokenizer.from_pretrained("t5-small")
```

```
mdl = T5ForConditionalGeneration.from_pretrained("t5-small")

def text2text_paraphrase(sentence1,sentence2):
    inp1 = "mrpc sentence1: "+sentence1
    inp2 = "sentence2: "+sentence2
    combined_inp=inp1+" "+inp2
    enc = text2text_tkn(combined_inp, return_tensors="pt")
    tokens = mdl.generate(**enc)
    response=text2text_tkn.batch_decode(tokens)
    return response

sent1=grad.Textbox(lines=1, label="Sentence1",
placeholder="Text in English")
sent2=grad.Textbox(lines=1, label="Sentence2",
placeholder="Text in English")
out=grad.Textbox(lines=1, label="Whether the sentence is
acceptable or not")
grad.Interface(text2text_paraphrase, inputs=[sent1,sent2],
outputs=out).launch()
```

requirements.txt
gradio
transformers
torch
transformers[sentencepiece]

Commit the changes and wait for the status of deployment to get green. Post that click the App tab in the menu to launch the application.

Provide inputs to the UI and click the Submit button to see the results as shown in Figure 5-42.

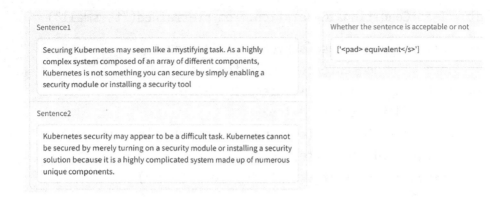

Figure 5-42. *Shows if sentences are equivalent or not*

Let's run the same code on two entirely different sentences in the following.

We get the following output.

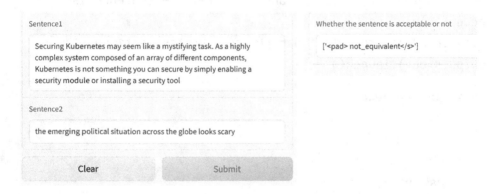

Figure 5-43. *Shows again if sentences are equivalent or not*

Next, we look into a task for checking whether a statement deduced from a text is correct or not. We again do this via the T5 model.

To achieve this we need to use the rte sentence1 sentence 2prefix as shown in the following code.

Code

app.py

Listing 5-27. app.py code

```python
from transformers import T5ForConditionalGeneration,
T5Tokenizer
import gradio as grad

text2text_tkn= T5Tokenizer.from_pretrained("t5-small")
mdl = T5ForConditionalGeneration.from_pretrained("t5-small")

def text2text_deductible(sentence1,sentence2):
    inp1 = "rte sentence1: "+sentence1
    inp2 = "sentence2: "+sentence2
    combined_inp=inp1+" "+inp2
    enc = text2text_tkn(combined_inp, return_tensors="pt")
    tokens = mdl.generate(**enc)
    response=text2text_tkn.batch_decode(tokens)
    return response

sent1=grad.Textbox(lines=1, label="Sentence1",
placeholder="Text in English")
sent2=grad.Textbox(lines=1, label="Sentence2",
placeholder="Text in English")
out=grad.Textbox(lines=1, label="Whether sentence2 is
deductible from sentence1")
grad.Interface(text2text_ deductible, inputs=[sent1,sent2],
outputs=out).launch()
```

> **requirements.txt**
> gradio
> transformers
> torch
> transformers[sentencepiece]

Commit the changes and wait for the status of deployment to get green. Post that click the App tab in the menu to launch the application.

Provide inputs to the UI and click the Submit button to see the results as shown in Figure 5-44.

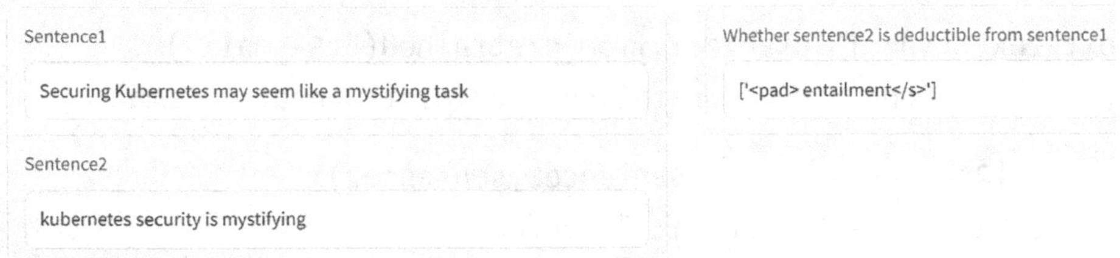

Sentence1	Whether sentence2 is deductible from sentence1
Securing Kubernetes may seem like a mystifying task	['<pad> entailment</s>']

Sentence2	
kubernetes security is mystifying	

Figure 5-44. *Gradio app for checking whether a sentence is deductible from another sentence or not – entailment*

Here, entailment means that sentence2 is deductible from sentence1.

Let's provide different sentences to the same task and see what we get as output.

This gives the following output.

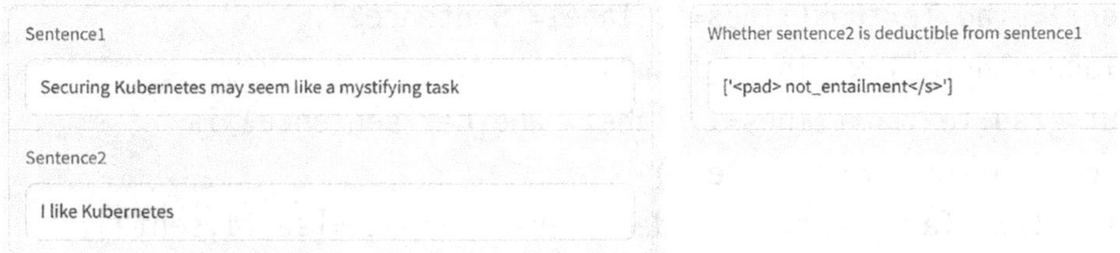

Sentence1	Whether sentence2 is deductible from sentence1
Securing Kubernetes may seem like a mystifying task	['<pad> not_entailment</s>']

Sentence2	
I like Kubernetes	

Figure 5-45. *Gradio app for checking whether a sentence is deductible from another sentence or not – not_entailment*

Here, not_entailment in the output signifies that sentence2 is not deductible from sentence1.

Moving away from T5 to the world of chatbots, we will show how easily we can develop a chatbot using huggingface APIs.

Chatbot/Dialog Bot

As a final example for this chapter, we take an example of how a simple dialog system can be built using the Transformers library.

Research in machine learning faces a formidable obstacle in the form of the construction of open-domain chatbots. While previous research has shown that scaling the neural models leads to improved results, this is not the only factor that should be considered when developing a good chatbot. A good conversation requires a lot of skills, which a chatbot needs to have to enter a seamless conversation.

These skills would entail understanding what is being conversed about and also what has been talked about in previous few sentences in the conversations. The bot should be able to handle scenarios where someone tries to trick it into out-of-context questions.

Below we showcase a simple bot named Alicia that is based on the Microsoft DialoGPT model.

Code

app.py

Listing 5-28. app.py code

```
from transformers import AutoModelForCausalLM, AutoTokenizer,Bl
enderbotForConditionalGeneration
import torch

chat_tkn = AutoTokenizer.from_pretrained("microsoft/
DialoGPT-medium")
mdl = AutoModelForCausalLM.from_pretrained("microsoft/
DialoGPT-medium")

#chat_tkn = AutoTokenizer.from_pretrained("facebook/
blenderbot-400M-distill")
```

```
#mdl = BlenderbotForConditionalGeneration.from_
pretrained("facebook/blenderbot-400M-distill")

def converse(user_input, chat_history=[]):

    user_input_ids = chat_tkn(user_input + chat_tkn.eos_token,
    return_tensors='pt').input_ids

    # keep history in the tensor
    bot_input_ids = torch.cat([torch.LongTensor(chat_history),
    user_input_ids], dim=-1)

    # get response
    chat_history = mdl.generate(bot_input_ids, max_length=1000,
    pad_token_id=chat_tkn.eos_token_id).tolist()
    print (chat_history)

    response = chat_tkn.decode(chat_history[0]).
    split("<|endoftext|>")

    print("starting to print response")
    print(response)

    # html for display
    html = "<div class='mybot'>"
    for x, mesg in enumerate(response):
        if x%2!=0 :
            mesg="Alicia:"+mesg
            clazz="alicia"
        else :
            clazz="user"

        print("value of x")
        print(x)
        print("message")
```

```
        print (mesg)

        html += "<div class='mesg {}'> {}</div>".
        format(clazz, mesg)
    html += "</div>"
    print(html)
    return html, chat_history

import gradio as grad

css = """
.mychat {display:flex;flex-direction:column}
.mesg {padding:5px;margin-bottom:5px;border-
radius:5px;width:75%}
.mesg.user {background-color:lightblue;color:white}
.mesg.alicia {background-color:orange;color:white,align-
self:self-end}
.footer {display:none !important}
"""
text=grad.inputs.Textbox(placeholder="Lets chat")
grad.Interface(fn=converse,
            theme="default",
            inputs=[text, "state"],
            outputs=["html", "state"],
            css=css).launch()
```

requirements.txt
gradio
transformers
torch
transformers[sentencepiece]

Commit the changes and wait for the status of deployment to get green. Post that click the App tab in the menu to launch the application.

Provide inputs to the UI and click the Submit button to see the conversation as shown in Figure 5-46.

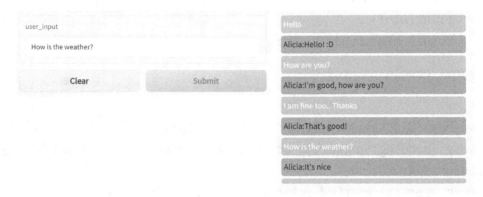

Figure 5-46. *Chatbot using the Microsoft DialoGPT model*

Code and Code Comment Generation

Code generation is quickly becoming a hot topic in the area of natural language processing (NLP), and contrary to popular belief, this is not just hype. The OpenAI Codex model was just just made available. If you view one of the Codex demonstrations, you will be able to observe how these models will influence the development of software programming in the future. Since the pretrained models are not made available to the public, working with a Codex model can be impossible from the point of view of a researcher if the requirements go beyond simply experimenting with it using the API. Technically, it is possible to recreate Codex by using the published paper; but, in order to do so, you will need a big GPU cluster, which is something that very few people either have access to or can afford. I believe that this restriction will make it more difficult to do research. Imagine if the authors chose not to disclose the pretrained weights; the number of BERT downstream applications that would be available to us would plummet. We can only hope that Codex is not the only paradigm of code generation currently available.

In this chapter, we will get an introduction to CodeGen, an encoder-decoder code generation model that features publicly available pretraining checkpoints that you can test out right now.

CodeGen

CodeGen is a language model that converts basic English prompts into code that can be executed. Instead of writing code yourself, you describe what the code should do using natural language, and the machine writes the code for you based on what you've described.

In the paper "A Conversational Paradigm for Program Synthesis," written by Erik Nijkamp, Bo Pang, Hiroaki Hayashi, Lifu Tu, Huan Wang, Yingbo Zhou, Silvio Savarese, and Caiming Xiong, there is a family of autoregressive language models for program synthesis called CodeGen. The models were first made available for download from this repository in three different pretraining data variations (NL, Multi, and Mono), as well as four different model size variants (350M, 2B, 6B, 16B).

In this chapter we look into some examples, where we can use the CodeGen model to generate code.

We will use a small CodeGen model with 350 million parameters.

Code

app.py

Listing 5-29. app.py code

```
from transformers import AutoTokenizer, AutoModelForCausalLM
import gradio as grad
codegen_tkn = AutoTokenizer.from_pretrained("Salesforce/
codegen-350M-mono")
```

```
mdl = AutoModelForCausalLM.from_pretrained("Salesforce/
codegen-350M-mono")

def codegen(intent):
# give input as text which reflects intent of the program.
    #text = " write a function which takes 2 numbers as input
    and returns the larger of the two"
    input_ids = codegen_tkn(intent, return_tensors="pt").
    input_ids

    gen_ids = mdl.generate(input_ids, max_length=128)
    response = codegen_tkn.decode(gen_ids[0], skip_special_
    tokens=True)
    return response

output=grad.Textbox(lines=1, label="Generated Python Code",
placeholder="")
inp=grad.Textbox(lines=1, label="Place your intent here")
grad.Interface(codegen, inputs=inp, outputs=output).launch()
```

requirements.txt
gradio
git+https://github.com/huggingface/transformers.git
torch
transformers[sentencepiece]

Commit the changes and wait for the status of deployment to get green. Post that click the App tab in the menu to launch the application.

Provide inputs to the UI and click the Submit button to see the results as shown in Figure 5-47.

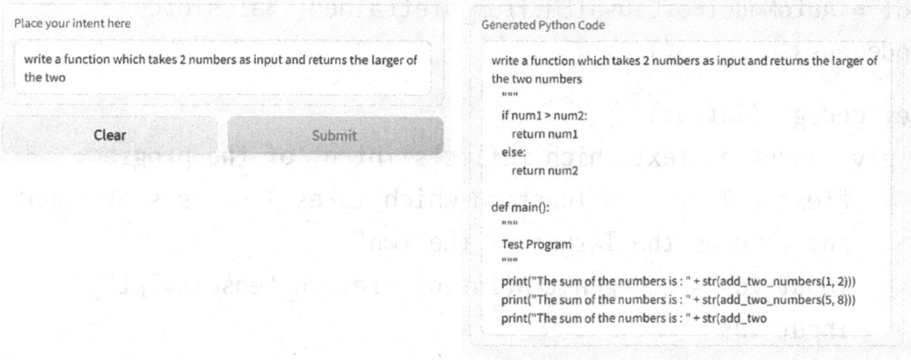

Figure 5-47. *Sample code generator for finding the larger of two numbers*

As we can see, the code is not fully accurate but captures the intent. Next, we try to generate code for bubble sort as shown in Figure 5-48.

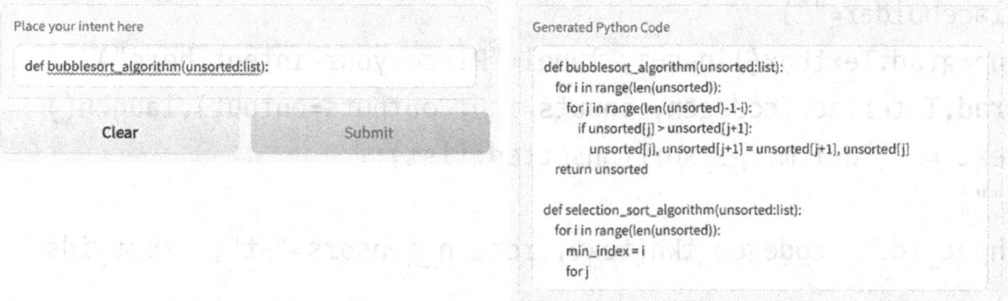

Figure 5-48. *Bubble sort text-to-code example*

The same can be done for merge sort by modifying the app.py file (change the max_length parameter to 256).

Listing 5-30. app.py code

```
from transformers import AutoTokenizer, AutoModelForCausalLM
import gradio as grad
codegen_tkn = AutoTokenizer.from_pretrained("Salesforce/
codegen-350M-mono")
```

```
mdl = AutoModelForCausalLM.from_pretrained("Salesforce/
codegen-350M-mono")

def codegen(intent):
# give input as text which reflects intent of the program.
    #text = " write a function which takes 2 numbers as input
    and returns the larger of the two"
    input_ids = codegen_tkn(intent, return_tensors="pt").
    input_ids

    gen_ids = mdl.generate(input_ids, max_length=256)
    response = codegen_tkn.decode(gen_ids[0], skip_special_
    tokens=True)
    return response

output=grad.Textbox(lines=1, label="Generated Python Code",
placeholder="")
inp=grad.Textbox(lines=1, label="Place your intent here")
grad.Interface(codegen, inputs=inp, outputs=output).launch()
text = """def merge_sort(unsorted:list):
"""

input_ids = codegen_tkn(text, return_tensors="pt").input_ids

gen_ids = mdl.generate(input_ids, max_length=256)
print(codegen_tkn.decode(gen_ids[0], skip_special_tokens=True))
```

Commit the changes and wait for the status of deployment to get green. Post that click the App tab in the menu to launch the application.

Provide inputs to the UI and click the Submit button to see the results as shown in Figure 5-49.

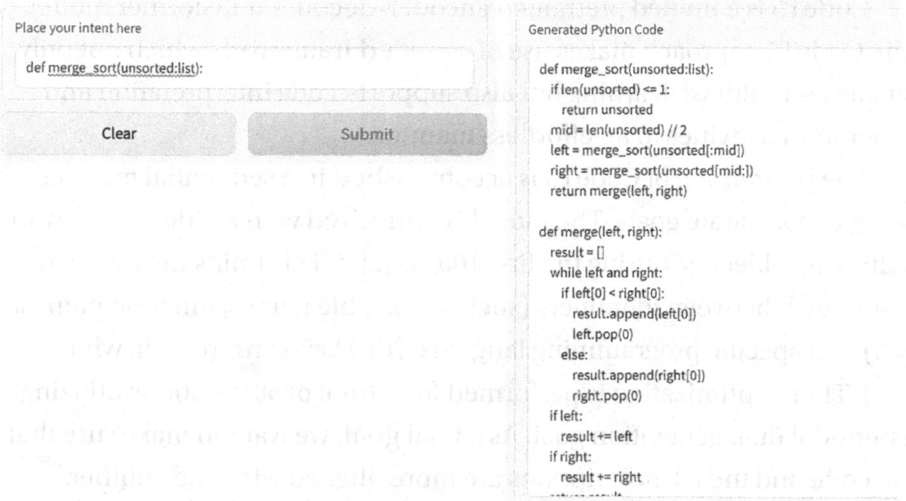

Figure 5-49. *Generating code for merge sort*

Code Comment Generator

The goal of this section is to have some code as input and let the model generate comments for that code. In this case we will use the Salesforce CodeT5 model, which is fine-tuned on Java code.

As its name suggests, the T5 encoder-decoder paradigm is the foundation upon which CodeT5 [1] is built. Instead of treating the source code like any other natural language (NL) text, it applies a new identifier-aware pretraining objective that capitalizes on code semantics. This is in contrast with previous code generation models, which rely on traditional pretraining methods.

The authors distributed two pretrained models: a basic model with 220 million data points and a smaller model with only 60 million data points. In addition to that, they distributed all of their fine-tuning checkpoints through their public GCP bucket. Additionally, the well-known huggingface library makes both of these pretrained models available for use.

CodeT5 is a unified pretrained encoder-decoder transformer model. The CodeT5 approach makes use of a unified framework, which not only facilitates multitask learning but also supports code interpretation and generation activities in an effortless manner.

The pretraining of CodeT5 is accomplished in a sequential manner using two separate goals. The model is optimized with an identifier-aware denoising objective during the first 100 epochs. This trains the model to distinguish between identifiers (such as variable names, function names, etc.) and specific programming language (PL) keywords (e.g., if, while, etc.). Then, optimization is performed for a total of 50 iterations utilizing a bimodal dual generation goal. As a final goal, we want to make sure that the code and the NL descriptions are more aligned with one another.

Since this example needs to download models from a non-huggingface repository (as of writing this book, the model was not updated on huggingface), we will do this example in Google Colab instead of huggingface.

Create a new notebook in Colab.

Before starting the comment generation code, we need to install the dependencies:

```
!pip install -q git+https://github.com/huggingface/
transformers.git
```

Create a comment model directory:

```
!mkdir comment_model
%cd comment_model
!wget -O config.json https://storage.googleapis.com/sfr-codet5-
data-research/pretrained_models/codet5_base/config.json

!wget -O pytorch_model.bin https://storage.googleapis.com/sfr-
codet5-data-research/finetuned_models/summarize_java_codet5_
base.bin
```

```
from transformers import RobertaTokenizer,
T5ForConditionalGeneration

model_name_or_path = './comment_model'  # Path to the folder
                                          created earlier.

codeT5_tkn = RobertaTokenizer.from_pretrained('Salesforce/
codet5-base')
mdl = T5ForConditionalGeneration.from_pretrained(model_name_
or_path)

    # provide code snippet as input
```

Listing 5-31. Code for comment generation from the source
code file

```
text = """ public static void main(String[] args) {

    int num = 29;
    boolean flag = false;
    for (int i = 2; i <= num / 2; ++i) {
      // condition for nonprime number
      if (num % i == 0) {
        flag = true;
        break;
      }
    }

    if (!flag)
      System.out.println(num + " is a prime number.");
    else
      System.out.println(num + " is not a prime number.");
  } """
```

```
input_ids = codeT5_tkn(text, return_tensors="pt").input_ids
gen_ids = mdl.generate(input_ids, max_length=20)

print(codeT5_tkn.decode(gen_ids[0], skip_special_tokens=True))
```

We get the following output:

```
A test program to check if the number is a prime number.
```

We input another text and check:

```
text = """ LocalDate localDate = new LocalDate(2020, 1, 31);
int numberOfDays = Days.daysBetween(localDate, localDate.
plusYears(1)).getDays();

boolean isLeapYear = (numberOfDays > 365) ? true : false;"""

input_ids = codeT5_tkn(text, return_tensors="pt").input_ids
gen_ids = mdl.generate(input_ids, max_length=150)

print(codeT5_tkn.decode(gen_ids[0], skip_special_tokens=True))
```

Output:

```
Returns true if the year is a leap year
```

Next, we try with a code that accesses the Google search APIs.

Listing 5-32. Code that tries to generate comment for Google search code

```
text = """
String google = "http://ajax.googleapis.com/ajax/services/
search/web?v=1.0&q=";
    String search = "stackoverflow";
    String charset = "UTF-8";
```

```
URL url = new URL(google + URLEncoder.encode(search,
charset));
Reader reader = new InputStreamReader(url.openStream(),
charset);
GoogleResults results = new Gson().fromJson(reader,
GoogleResults.class);

// Show title and URL of 1st result.
System.out.println(results.getResponseData().getResults().
get(0).getTitle());
System.out.println(results.getResponseData().getResults().
get(0).getUrl());
"""
```

```
input_ids = codeT5_tkn(text, return_tensors="pt").input_ids
gen_ids = mdl.generate(input_ids, max_length=50,
temperature=0.2,num_beams=200,no_repeat_ngram_size=2,num_
return_sequences=5)
```

```
print(codeT5_tkn.decode(gen_ids[0], skip_special_tokens=True))
```

This outputs

```
https://www. googleapis. com / ajax. services. search. web?v =
1. 0 &q = 123 Show title and URL of 1st result.
```

The last result might not look good, but this can be improved by tuning the specific parameters, which I leave to you to experiment with.

Finally, these pretrained models can also be fine-tuned for specific programming languages like C, C++, etc.

Summary

In this chapter we looked into the various use cases and implementations of how transformers can be used for processing text and applying it for various tasks like classification, translation, summarization, etc. We also looked at how we can easily build a chatbot using the huggingface APIs.

In the next chapter, we look at how transformers can be applied to the area of image processing.

CHAPTER 6

Fine-Tuning Pretrained Models

So far we have seen how to use huggingface APIs that include the pretrained models to create simple applications. Wouldn't it be amazing if you could start from scratch and train your own model using only your own data?

Utilizing transfer learning is the most effective strategy to take if you do not have a large amount of spare time or computing resources at your disposal. There are two main advantages of utilizing transfer learning with Hugging Face as opposed to starting from scratch when training a model.

As we stated in Chapter 4, models like GPT3 take enormous amount of infrastructural resources to train. This is beyond the capability of most of us. So how do we then use these models in a much more flexible way and not just use them by downloading the pretrained models? The answer lies in fine-tuning these models with the additional data we have. This would require very little resources and be easy to achieve as compared with training a full big language model from scratch.

To transform a basic model into something that is able to generate reliable outcomes requires a significant investment of time and resources. Because of transfer learning, you can forego the laborious step of training and devote only a small amount of time to adapting the dataset to your specific requirements.

© Shashank Mohan Jain 2022
S. M. Jain, *Introduction to Transformers for NLP*,
https://doi.org/10.1007/978-1-4842-8844-3_6

In fact, pretrained models from Hugging Face are capable of excelling at tasks in a variety of domains even without the need for additional fine-tuning. It is likely that one can use these models in a zero-shot learning scenario as well, but if there is a specific dataset one has, then our good friend, the huggingface APIs, provides us the needed abstractions for fine-tuning these existing models.

Therefore, we can basically consider transfer learning to be a kind of a shortcut when it comes to training. Simply by making use of pretrained language models, you can save tens of thousands of dollars and thousands of hours in terms of your computing needs. You should stick to transfer learning unless the tasks you are working on are extremely specific and cannot be solved using models that already exist.

We can now move on to our fine-tuning guide with Hugging Face because we have a better understanding of the applications and advantages of transfer learning.

The workflow for fine-tuning is shown in the following:

- Select a pretrained model from huggingface that suits the need for your use case.

- The additional custom dataset has to stick to the huggingface dataset spec, so we need to preprocess our data so that it conforms to the format needed.

- Upload the dataset to Colab, S3, or any other store.

- Use the Trainer API from huggingface to fine-tune the existing model.

- Save the model locally or upload it to the huggingface repo.

With some basic idea, let's get started on some transfer learning with Hugging Face libraries.

During the fine-tuning stages, most of the neural architecture is frozen. This means that we only adjust weights in the output layer. Since we have already covered tokenizers in an earlier chapter, we will give a brief overview about huggingface datasets here, which are the most important constructs for this chapter. Once we understand the Datasets API, we will proceed to using a custom dataset for a pretrained model through transfer learning.

Datasets

In this section we describe the basic dataset construct from huggingface and some of its basic functions.

The data that you are utilizing throughout the course of any machine learning project is going to be of utmost significance. The real accuracy derives not only from the quantity but also from the quality of the data that is being used, and this is true regardless of the algorithm or model that you are working with.

Accessing large datasets can be a challenging endeavor at times. The process of scraping, accumulating, and then cleaning up this data in the appropriate manner can take a significant amount of time. Hugging Face, fortunately for people interested in NLP as well as image and audio processing, comes with a central repository of datasets that are already prepared for use. In the following paragraphs, we will have a brief look at how you can work with this datasets module to select and prepare the appropriate dataset for your project.

To install the datasets library, use the following command:

```
!pip install datasets
```

As we have been reading the documentation for the datasets repository, we have discovered that there are several primary methods. The first method is the one that we are able to use to investigate the list of datasets that are readily available. You should see options to work with close to 6800 different datasets, all of which are currently available:

```
from datasets import list_datasets, load_dataset, list_metrics,
load_metric

# Print all the available datasets
print(len(list_datasets()))
```

Output:

```
6783
```

Load a dataset:

```
dataset = load_dataset('imdb')
```

Print the dataset object:

```
DatasetDict({ train: Dataset({ features: ['text', 'label'],
num_rows: 25000 }) test: Dataset({ features: ['text', 'label'],
num_rows: 25000 }) unsupervised: Dataset({ features: ['text',
'label'], num_rows: 50000 }) })
```

It constitutes a dictionary with train, test, and unsupervised datasets, each having features and num_rows as values. Here, examples are taken from the IMDB dataset, and thereby the text on which we will be doing sentiment analysis is also taken from IMDB.

Let us access the train dataset:

```
dataset['train'][2]
```

```
{'label': 0, 'text': "If only to avoid making this type of film
in the future. This film is interesting as an experiment but
tells no cogent story.<br /><br />One might feel virtuous for
sitting thru it because it touches on so many IMPORTANT issues
but it does so without any discernable motive. The viewer comes
away with no new perspectives (unless one comes up with one
while one's mind wanders, as it will invariably do during this
pointless film).<br /><br />One might better spend one's time
staring out a window at a tree growing.<br /><br />"}
```

Describe the dataset:

```
dataset['train'].description
```

We get the following output:

```
Large Movie Review Dataset.\nThis is a dataset for binary
sentiment classification containing substantially more data
than previous benchmark datasets. We provide a set of 25,000
highly polar movie reviews for training, and 25,000 for
testing. There is additional unlabeled data for use as well.
```

List features of the dataset:

```
dataset['train'].features
```

We can see there are two features:

```
{'label': ClassLabel(num_classes=2, names=['neg', 'pos'],
id=None), 'text': Value(dtype='string', id=None)}
```

It's possible that you won't want to deal with utilizing one of the Hugging Face datasets in certain circumstances. This dataset object is still capable of loading locally stored CSV files in addition to other types of files. If, for example, you want to work with a CSV file, you can easily pass this information into the load dataset method along with the path to the CSV file on your local machine.

Fine-Tuning a Pretrained Model

Now since we understand the dataset construct, it's time for applying some transfer learning on a pretrained model using our own dataset. In the following we show an example of how to fine-tune a pretrained model with the IMDB dataset.

We will divide the fine-tuning aspect into two parts. The training section is where we will use the Trainer API of huggingface to fine-tune the model and save it. The other section is the inference part where we will load this fine-tuned model to achieve inferencing.

Training for Fine-Tuning

First, install transformers and datasets using the following command:

```
!pip install datasets transformers
```

Next, load the IMDB dataset:

```
from datasets import load_dataset
dataset = load_dataset("imdb")
dataset["train"][100]
```

The following is a sample review:

```
{'label': 0, 'text': "Terrible movie. Nuff Said.<br /><br
/>These Lines are Just Filler. The movie was bad. Why I have
to expand on that I don't know. This is already a waste of
my time. I just wanted to warn others. Avoid this movie. The
acting sucks and the writing is just moronic. Bad in every way.
Even that was ruined though by a terrible and unneeded rape
scene. The movie is a poorly contrived and totally unbelievable
piece of garbage.<br /><br />OK now I am just going to rag on
IMDb for this stupid rule of 10 lines of text minimum. First I
```

waste my time watching this offal. Then feeling compelled to warn others I create an account with IMDb only to discover that I have to write a friggen essay on the film just to express how bad I think it is. Totally unnecessary."}

Next, we need to tokenize the dataset we loaded using the BERT tokenizer. The first step is to create a new Jupyter notebook in Google Colab and copy the following code line by line:

```
from transformers import AutoTokenizer

brt_tkn = AutoTokenizer.from_pretrained("bert-base-cased")

def generate_tokens_for_imdb(examples):
    return brt_tkn(examples["text"], padding="max_length",
    truncation=True)

tkn_datasets = dataset.map(generate_tokens_for_imdb,
batched=True)
```

The aforementioned code yields the following output:

```
loading configuration file https://huggingface.co/bert-base-
cased/resolve/main/config.json from cache at /root/.cache/
huggingface/transformers/a803e0468a8fe090683bdc453f4fac622804
f49de86d7cecaee92365d4a0f829.a64a22196690e0e82ead56f388a3ef
3a50de93335926ccfa20610217db589307
Model config BertConfig {
  "_name_or_path": "bert-base-cased",
  "architectures": [
    "BertForMaskedLM"
  ],
  "attention_probs_dropout_prob": 0.1,
  "classifier_dropout": null,
```

```
  "gradient_checkpointing": false,
  "hidden_act": "gelu",
  "hidden_dropout_prob": 0.1,
  "hidden_size": 768,
  "initializer_range": 0.02,
  "intermediate_size": 3072,
  "layer_norm_eps": 1e-12,
  "max_position_embeddings": 512,
  "model_type": "bert",
  "num_attention_heads": 12,
  "num_hidden_layers": 12,
  "pad_token_id": 0,
  "position_embedding_type": "absolute",
  "transformers_version": "4.20.1",
  "type_vocab_size": 2,
  "use_cache": true,
  "vocab_size": 28996
}
```

```
loading file https://huggingface.co/bert-base-cased/resolve/
main/vocab.txt from cache at /root/.cache/huggingface/
transformers/6508e60ab3c1200bffa26c95f4b58ac6b6d95fba4db1f
195f632fa3cd7bc64cc.437aa611e89f6fc6675a049d2b5545390adbc617
e7d655286421c191d2be2791
loading file https://huggingface.co/bert-base-cased/resolve/
main/tokenizer.json from cache at /root/.cache/huggingface/
transformers/226a307193a9f4344264cdc76a12988448a25345ba172f
2c7421f3b6810fddad.3dab63143af66769bbb35e3811f75f7e16b2320e
12b7935e216bd6159ce6d9a6
loading file https://huggingface.co/bert-base-cased/resolve/
main/added_tokens.json from cache at None
```

```
loading file https://huggingface.co/bert-base-cased/resolve/
main/special_tokens_map.json from cache at None
loading file https://huggingface.co/bert-base-cased/resolve/
main/tokenizer_config.json from cache at /root/.cache/
huggingface/transformers/ec84e86ee39bfe112543192cf981deebf7e
6cbe8c91b8f7f8f63c9be44366158.ec5c189f89475aac7d8cbd243960a06
55cfadc3d0474da8ff2ed0bf1699c2a5f
loading configuration file https://huggingface.co/bert-base-
cased/resolve/main/config.json from cache at /root/.cache/
huggingface/transformers/a803e0468a8fe090683bdc453f4fac622804
f49de86d7cecaee92365d4a0f829.a64a22196690e0e82ead56f388a3ef3
a50de93335926ccfa20610217db589307
Model config BertConfig {
  "_name_or_path": "bert-base-cased",
  "architectures": [
    "BertForMaskedLM"
  ],
  "attention_probs_dropout_prob": 0.1,
  "classifier_dropout": null,
  "gradient_checkpointing": false,
  "hidden_act": "gelu",
  "hidden_dropout_prob": 0.1,
  "hidden_size": 768,
  "initializer_range": 0.02,
  "intermediate_size": 3072,
  "layer_norm_eps": 1e-12,
  "max_position_embeddings": 512,
  "model_type": "bert",
  "num_attention_heads": 12,
  "num_hidden_layers": 12,
  "pad_token_id": 0,
```

```
"position_embedding_type": "absolute",
"transformers_version": "4.20.1",
"type_vocab_size": 2,
"use_cache": true,
"vocab_size": 28996
}
```

Once we tokenize the dataset, we will only be fine-tuning on 200 samples, so that we can tune the model faster for simplicity's sake. You are encouraged to try with more samples:

```
training_dataset = tokenized_datasets["train"].
shuffle(seed=42).select(range(200))
evaluation_dataset = tokenized_datasets["test"].
shuffle(seed=42).select(range(200))
```

Load the BERT-based sequence classification model:

```
from transformers import AutoModelForSequenceClassification

mdl = AutoModelForSequenceClassification.from_pretrained("bert-base-cased", num_labels=2)
```

The Transformers library includes a Trainer class that is specifically designed for training huggingface transformer models. This class makes it much simpler to begin training without the need to manually write your own code. The Trainer API provides features like logging, monitoring, etc.

Here, we provide the training arguments by instantiating a class called TrainingArguments that has all of the hyperparameters that one can experiment with. Here in this case, we will be just using the defaults:

```
from transformers import TrainingArguments

training_args = TrainingArguments(output_dir="imdb")
```

During training, Trainer does not automatically evaluate how well the model is performing. If you want Trainer to be able to compute and report metrics, you will need to pass it a function. This is what we will do in the following code segment:

```
import numpy as np
from datasets import load_metric

mdl_metrics = load_metric("accuracy")

def calculate_metrics(eval_pred):
    logits, labels = eval_pred
    predictions = np.argmax(logits, axis=-1)
    return mdl_metrics.compute(predictions=predictions,
    references=labels)
```

```
from transformers import TrainingArguments, Trainer
```

```
trng_args = TrainingArguments(output_dir="test_trainer",
evaluation_strategy="epoch", num_train_epochs=3)
```

Instantiate a Trainer object that contains your model, the training arguments, the datasets to be used for training and testing, and the evaluation function:

```
Mdl_trainer = Trainer(
    model=model,
    args=trng_args,
    train_dataset=training_dataset,
    eval_dataset=evaluation_dataset,
    compute_metrics=calculate_metrics,
)
```

Train the model:

```
trainer.train()
```

Figure 6-1 shows the training run for the IMDB dataset we used for fine-tuning an existing pretrained model.

```
***** Running training *****
  Num examples = 200
  Num Epochs = 3
  Instantaneous batch size per device = 8
  Total train batch size (w. parallel, distributed & accumulation) = 8
  Gradient Accumulation steps = 1
  Total optimization steps = 75
```
[75/75 01:17, Epoch 3/3]

Epoch	Training Loss	Validation Loss	Accuracy
1	No log	0.654922	0.590000
2	No log	0.504451	0.780000
3	No log	0.545163	0.800000

```
The following columns in the evaluation set don't have a corresponding argument in
***** Running Evaluation *****
  Num examples = 200
  Batch size = 8
The following columns in the evaluation set don't have a corresponding argument in
***** Running Evaluation *****
  Num examples = 200
  Batch size = 8
The following columns in the evaluation set don't have a corresponding argument in
***** Running Evaluation *****
  Num examples = 200
  Batch size = 8
```

Figure 6-1. *Training run for the IMDB dataset for fine-tuning*

Save the fine-tuned trained model:

```
trainer.save_model()
```

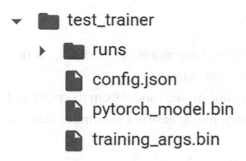

Figure 6-2. *Save the model locally (we have a PyTorch-based model with extension .bin)*

We can see that the fine-tuned model is saved with the name pytorch_model.bin.

We can check the accuracy of the model using the following code:

```
metrics = mdl_trainer.evaluate(evaluation_dataset)
trainer.log_metrics("eval", metrics)
trainer.save_metrics("eval", metrics)
```

```
***** Running Evaluation *****
  Num examples = 200
  Batch size = 8
                                               [25/25 00:06]
***** eval metrics *****
  eval_accuracy             =            0.8
  eval_loss                 =         0.5452
  eval_runtime              =      0:00:06.36
  eval_samples_per_second   =         31.401
  eval_steps_per_second     =          3.925
```

Figure 6-3. *Evaluation of the fine-tuned model in terms of its accuracy*

Inference

Once we have fine-tuned the model and saved it, it's time to do inference on data outside the train dataset.

We will load the fine-tuned model from the path and use it to make a classification, which in this case is a sentiment classification on IMDB movie reviews:

```
from transformers import BertTokenizer
```

Load the fine-tuned model from the following path:

```
PATH = 'test_trainer/'
md = AutoModelForSequenceClassification.from_pretrained(PATH,
local_files_only=True)

def make_classification(text):
    # Tokenize
    inps = brt_tkn(text, padding=True, truncation=True,
    max_length=512, return_tensors="pt").to("cuda")
    # get output
    outputs = model(**inps)
    # softmax for generating probablities
    probablities = outputs[0].softmax(1)
    # get best match.
    return probablities .argmax()
```

Here is the first inference:

```
text = """
This is the show that puts a smile on your face as you watch
it. You get in love with each and every character of the show.
At the end, I felt eight episode were not enough. Will wait for
season 2.
"""
```

```
print(make_classification(text))
```

This yields the following output:

```
tensor(1, device='cuda:0')
Output of 1 is positive review
```

Here is the second inference:

```
text = """
It was fun to watch but It did not impress that much I think i
waste my money popcorn time pizza burgers everything.

Akshay should make only comedy movies these King type movies
suits on king like personality of actors Total waste.
"""

print(make_classification(text))
```

This yields the following output:

```
tensor(0, device='cuda:0')
Output of zero is negative review
```

Summary

In this chapter, we learned about the huggingface datasets and their different functions. We also learned how we can use the huggingface APIs for fine-tuning existing pretrained models with other datasets.

APPENDIX A

Vision Transformers

Before the advent of vision transformers (ViTs), all tasks related to image-based machine learning like image classification, object detection, Q&A on images, image-to-caption mapping, etc. were taken care of by mostly CNNs and related neural architectures. With vision transformers there emerged an alternate means of handling such image-related tasks with better results.

Among the papers released on vision transformers, the one released on October 22, 2020, by Alexey Dosovitskiy, Lucas Beyer, Alexander Kolesnikov, Dirk Weissenborn, Xiaohua Zhai, and Thomas Unterthiner is one that's particularly noteworthy. Their approach is based on Vaswani's "Attention Is All You Need" paper, which is widely used in natural language processing and has been referred to in previous chapters. There were no changes made to the attention layers in this paper. Breaking an image into little patches (perhaps 16×16) is their most essential trick.

Self-Attention and Vision Transformers

How you apply self-attention to images is the big question. As in NLP where one word pays attention to other words (to find the relation between the words), we need to apply a similar concept to images. The important aspect to understand here is that how we achieve this mechanism. This is where vision transformers come into the picture.

© Shashank Mohan Jain 2022
S. M. Jain, *Introduction to Transformers for NLP*,
https://doi.org/10.1007/978-1-4842-8844-3

153

To achieve self-attention, vision transformers divide the image into different parts. Each part of the image is a linear sequence of vectors, which constitutes the pixel values. The only thing we do is that we have reduced the 2D representation of this part of the image into a 1D vector representation. Post this, to each of this representation a positional embedding is done, so that we have positional semblance maintained within the learned representation. This is similar in nature to the positional embedding we have seen for the text embeddings in previous chapters.

At a high level, the architecture of a vision transformer is shown in Figure A-1.

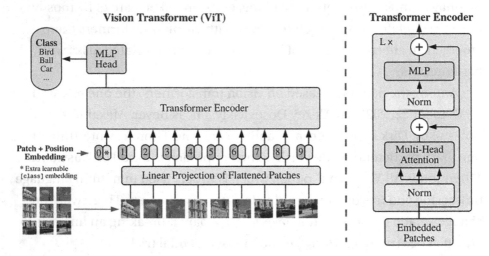

Figure A-1. *Architecture of a ViT taken from the paper*

The multi-layer perceptron (MLP) layer and the multi-headed self-attention (MSA) layer are both components of the transformer encoder module. The multi-headed self-attention layer divides the inputs into numerous heads, allowing each head to learn a distinct level of self-attention independently. After that, the outputs of each of the heads are stringed together, and then the multi-layer perceptron layer processes them.

The book is not about going into details of the vision transformer itself. The interested reader can seek further knowledge from the paper on vision transformers.

Before we look into the code, there is an important class called FeatureExtractor in the huggingface library.

In most cases, the job of preparing input features for models that don't fit within the traditional NLP models falls on the shoulders of a feature extractor. They are responsible for a variety of tasks, including the manipulation of photographs and the processing of audio recordings. The majority of vision products are packaged with an additional feature extractor.

Without going into details of different aspects of vision transformers (as NLP is the main focus of this book), we illustrate one example for image classification via vision transformers.

We illustrate in the following the code for using a ViT for an image classification task.

As with code samples in Chapter 5, we use Gradio as the framework here, and so our example below will follow the same pattern that we adopted in Chapter 5.

Image Classification Using a ViT

Code

app.py

Listing A-1. Gradio app for vision transformers

```
import gradio as grad
grad.Interface.load(
            "models/microsoft/swin-tiny-patch4-window7-224",
            theme="default",
            css=".footer{display:none !important}",
            title=None).launch()
```

Here, the gr.Interface.load method loads the model we provided as the path. In this example it is models/microsoft/swin-tiny-patch4-window7-224 .

We get output as shown in Figure A-2.

Figure A-2. *Gradio app that takes an image of a cat and classifies it into a breed of a cat*

Uploading a dog image classifies it properly again, as shown in Figure A-3.

Figure A-3. *Gradio app that takes an image of a dog and classifies it into a breed of a dog*

There are other tasks like image segmentation and object detection possible via vision transformers. We leave this to you to try out as an exercise.

Summary

In this Appendix, we touched upon vision transformers in a very brief manner. The intent here was not to take a deep dive into vision transformers as this book is mainly around NLP. But since this is also a new development in the area of transformers, the author thinks that having knowledge of this area will be handy for the readers.

We will discuss how we can use huggingface for fine-tuning purposes.

Index

A

Application programming interface (API), 52–54, 126, 134

Artificial general intelligence (AGI), 1, 53

Artificial intelligence (AI), 1, 2, 16, 17, 49, 51

Attention, 14, 15, 27, 31–33, 35, 57, 153

"Attention Is All You Need", 19, 21, 23, 51, 153

Attention mechanism, 15, 29, 30, 36, 38, 51

AutoModel
    probabilities, 65, 66
    text classification, 63
    Transformers library, 63, 64
    wrapper class, 66, 67

AutoTokenizer, 56–58, 60–63, 66, 73, 77, 79, 80, 83, 107, 109, 123, 127, 129, 143

B

Backpropagation (BP), 9, 11

Bag of words, 4–6

BERT-Base, 37, 44, 52

BERT-based Q&A, 78

BERT-based tokenizer, 58

BERT-Large, 37, 44, 52

Bidirectional Encoder Representations from Transformers (BERT), 41, 42, 52
    application, bidirectional training, 37
    BERT-Base, 37
    BERT-Large, 37
    huggingface, 45
    inference in NSP, 43, 44
    input representations, 45, 46
    LSTM-type architectures, 38
    MLM, 37, 38, 40, 41
    performance, 48, 49
    pretrained models, 44
    sentiment analysis, tweet, 47, 48
    training language models process, 38
    transfer learning, 37
    transformer architecture, 38
    use cases, 46, 47

C

Chatbot/dialog bot, 123–125

CLS and SEP, 59

© Shashank Mohan Jain 2022
S. M. Jain, *Introduction to Transformers for NLP*,
https://doi.org/10.1007/978-1-4842-8844-3